Medical Affairs' Fundamentals: The Next Generation

Andriy Krendyukov

Medical Affairs' Fundamentals: The Next Generation

Andriy Krendyukov
Chief Medical Officer
Aspire Pharma
Munich, Germany

ISBN 978-3-031-92587-0   ISBN 978-3-031-92588-7 (eBook)
https://doi.org/10.1007/978-3-031-92588-7

This Springer imprint is published by the registered company Springer Nature Switzerland AG
The registered company address is: Gewerbestrasse 11, 6330 Cham, Switzerland

Preface

Kurt Lewin once remarked, *"There is nothing so practical as a good theory."* This notion applies profoundly to career transitions in the pharmaceutical fields. A well-structured career trajectory, built on a strong scientific foundation and knowledge, leads not only to professional fulfillment but also to meaningful contributions toward patient care. The pharmaceutical industry provides a dynamic environment for healthcare professionals (HCPs) transitioning from clinical or research backgrounds, where their expertise is leveraged to advance medical innovation and improve patients' lives.

Navigating a Career Transition: The Underlying Motivations

The decision to transition to a career in the pharmaceutical industry is often influenced by a multitude of factors. Studies have consistently identified key drivers of career change, including financial incentives, professional growth, work-life balance, and a desire for broader impact.

Couple of years ago, just before embarking on the journey to write this book, the author conducted a LinkedIn survey, results of which confirmed the most common reasons professionals consider a move into the pharmaceutical sector:

- *Financial Incentives*: Competitive remuneration and comprehensive benefits
- *Recognition and Appreciation*: A work environment that values scientific and clinical expertise
- *Reduced Stress and Improved Work-Life Balance*: Opportunities to apply medical knowledge outside the high-intensity clinical setting
- *Career Advancement*: Roles that offer scientific and leadership growth
- *Commitment to Innovation*: Engaging in the development of novel treatments that transform patient outcomes

These motivations illustrate a common theme: professionals do not transition merely for a change in setting, but rather for the opportunity to contribute to health-care in an innovative and strategic manner.

Bridging Science and Strategy: The Role of Medical Affairs

For physicians, pharmacists, researchers, and other healthcare professionals, Medical Affairs offers a unique career path that merges scientific expertise with business acumen, regulatory understanding, and strategic planning. This function plays a crucial role in ensuring that innovative therapies reach patients safely and effectively while maintaining scientific integrity and regulatory compliance.

Core Functions of Medical Affairs

Medical Affairs professionals operate at the intersection of science, medicine, and business, with responsibilities including:

- *Evidence Generation*: Designing and overseeing post-marketing studies to fill clinical gaps
- *Scientific Communication*: Disseminating accurate, evidence-based information to healthcare professionals
- *Stakeholder Engagement*: Collaborating with regulatory bodies, patient advocacy groups, and key opinion leaders
- *Strategic Decision-Making*: Contributing to product development, commercialization strategies, and market access planning

This diverse scope of responsibilities highlights the indispensable role of Medical Affairs in ensuring that therapies are utilized optimally to improve patient outcomes.

From Clinical Practice to Pharmaceutical Innovation

Consider the journey of a practicing physician who transitioned from a university hospital to a leading pharmaceutical company. Initially, the physician thrived in a surgical role, gaining technical mastery and clinical acumen. However, over time, a sense of professional plateau emerged—despite the satisfaction of direct patient care, the desire to influence healthcare on a broader scale became evident.
A serendipitous encounter with a former colleague working in the pharmaceutical sector opened new doors. The physician's transition was further catalyzed by a period of forced reflection due to a temporary medical leave. This pause provided the opportunity to consider how their expertise could extend beyond the operating

room. With institutional support and a calculated risk-taking approach, the physician successfully navigated the transition and never returned to clinical practice, finding fulfillment in a role where science, strategy, and patient impact converged.

Essential Skills for Success in the Pharmaceutical Industry

For those considering a transition, success in the pharmaceutical sector requires a distinct set of competencies beyond clinical and research expertise. Key attributes include:

- *Scientific Acumen*: A deep understanding of disease mechanisms, therapeutic areas, and clinical research methodologies
- *Communication Skills*: The ability to translate complex medical data into actionable insights for diverse stakeholders
- *Adaptability*: Navigating the ever-evolving landscape of regulatory requirements, market dynamics, and scientific advancements
- *Ethical Integrity*: Upholding patient-centricity and regulatory compliance in all decision-making processes
- *Strategic Thinking*: Integrating scientific knowledge with business objectives to drive medical innovation and patient access

The Patient-Centric Focus of Pharmaceutical Careers

Regardless of functional areas, one fundamental principle unites all pharmaceutical industry professionals: patient-centricity. Every role, every strategy, and every piece of scientific communication is ultimately designed to enhance patient care and improve quality of life.
The late the founder of the top ten pharmaceutical company aptly captured in this mission statement: "*There is so much more to be done; the patients are waiting*." This sentiment serves as a reminder that beyond business considerations, the true driving force behind pharmaceutical careers is the well-being of patients worldwide.

Encouragement for Aspiring Professionals

For those contemplating a move to the pharmaceutical industry, the path may seem unfamiliar at first, but it is rich with opportunity. Transitioning from clinical or research roles to industry positions is not a deviation from patient care but rather an evolution of it. By leveraging their expertise, healthcare professionals can

contribute to advancements in medical science, regulatory frameworks, and healthcare accessibility on a global scale.
The pharmaceutical industry welcomes professionals who are eager to drive innovation while upholding the highest standards of scientific integrity. Whether in Medical Affairs, regulatory sciences, or research and development, each role contributes to the shared mission of delivering life-changing treatments to patients in need.

To summarize

Career transitions are complex, often requiring deep reflection, strategic planning, and the courage to embrace change. For those driven by a passion for science, innovation, and patient care, the pharmaceutical industry offers a compelling path to expand their impact beyond traditional clinical settings.

As we look to the future, the opportunities within the industry will continue to grow, driven by advancements in precision medicine, digital health, and regulatory innovation. For aspiring professionals, the message is clear: *embrace the possibilities, seek continuous learning, and remain steadfast in your commitment to improving patients' lives.*

To all those embarking on this journey—welcome to a career where science meets strategy, and where every innovation serves a singular purpose: transforming patient care.

As you will be transitioning to the following chapters, this book will delve deeper into the multifaceted landscape of Medical Affairs. These sections will provide a comprehensive overview of its core functions, essential skills, and potential career trajectories, offering aspiring professionals the insights needed to navigate and thrive in this dynamic field.

Munich, Germany Andriy Krendyukov

Contents

Part I

Chapter 1
Medical Affairs: What It Is and What It Isn't

Key Highlights

- Evolution of the pharmaceutical industry from small molecules to biologics and personalized medicine.
- The expanding role of Medical Affairs from post-marketing support to strategic involvement across the product lifecycle.
- Increasing importance of RWE, patient-reported outcomes, and compliance with evolving regulations.
- Medical Affairs as the bridge between R&D, commercial teams, and external stakeholders, ensuring scientific and ethical integrity.
- Medical Affairs' contribution to shaping research priorities, product positioning, and market access strategies.

The pharmaceutical industry's history is one of transformative innovation, characterized by scientific breakthroughs and continuous evolution within a complex regulatory framework. From its early foundations, the pharmaceutical sector has undergone profound changes, contributing not only to the treatment of diseases but also to enhancing overall quality of life. The industry's advancements have shifted from treating acute conditions to addressing chronic diseases and even personalizing healthcare based on individual genetic profiles.

One of the pivotal moments in pharmaceutical history occurred in 1928, when Alexander Fleming discovered penicillin. This discovery marked the dawn of the antibiotic era, revolutionizing the treatment of infectious diseases and setting the stage for rapid developments in medical science. The decades following Fleming's discovery saw major innovations, including the development of the first vaccines in the 1940s, psychotropic medications in the 1950s, and the emergence of biotechnology in the late 1970s, leading to recombinant DNA technology. These technological leaps enlarged the scope of medicine, allowing the pharmaceutical industry to expand from producing small molecules to biologics and advanced therapies such as monoclonal antibodies, CAR-T cell therapies, and mRNA vaccines.

A. Krendyukov, *Medical Affairs' Fundamentals: The Next Generation*,
https://doi.org/10.1007/978-3-031-92588-7_1

The late twentieth century and early twenty-first century heralded a new age of personalized medicine, driven by advances in genetic research and the Human Genome Project. This shift from a "one-size-fits-all" approach to tailored therapeutic strategies underscored the importance of understanding individual genetic profiles to optimize treatment outcomes. The continuous growth in molecular biology and computational biology is propelling the pharmaceutical industry toward therapies that are not only more effective but also more specific to the needs of individual patients.

The Role of Medical Affairs

Historically, even though it emerged as a separate entity less than two decades ago, the role of Medical Affairs in pharmaceutical companies was largely supportive, acting as a liaison between clinical development teams and the commercial divisions. In its early years, the department was primarily involved in post-marketing activities, such as providing medical information and supporting product launches. Within the last decade, the role of Medical Affairs has significantly evolved, becoming a critical pillar in strategic decision-making processes within pharmaceutical companies.

This evolution has been largely driven by the increasing complexity of medicine development, the growing demand for scientific rigor, and the necessity to comply with ever-tightening regulatory standards. Medical Affairs now plays a vital role across the entire product lifecycle—from the preclinical and early stages of clinical development to post-market surveillance. Its responsibilities include designing clinical programs and trials, educating healthcare professionals, engaging with external stakeholders, and ensuring compliance with ethical and regulatory frameworks.

Today, Medical Affairs should serve as the scientific and strategic backbone of pharmaceutical companies, bridging the gap between R&D, internal and external stakeholders, and the healthcare system. The department is essential for ensuring that the medicinal products or medical devices are developed to be clinically sound and to align with the needs of patients, HCPs and providers, and healthcare systems globally. By integrating clinical research and evidence with commercial strategies, Medical Affairs helps to ensure that pharmaceutical companies remain competitive while maintaining a strong ethical foundation.

Changing Landscape of Healthcare

The healthcare landscape is undergoing continuous and profound changes, driven by technological advancements, shifting patient demographics, and evolving regulatory environments. These dynamics demand that every facet of healthcare, including pharmaceutical companies and their Medical Affairs functions, adapt to new realities.

One of the most significant transformations in healthcare has been the rise of digital health technologies. Innovations such as electronic health records (EHRs), telemedicine, and mobile health applications have altered how healthcare is delivered and accessed. These technologies enable more efficient data collection, improved patient outcomes, and better patient-provider communication. As a result, Medical Affairs must now place a greater emphasis on RWE and patient-reported outcomes, employing data analytics to inform decision-making processes and ensure that products deliver measurable benefits in real-world settings.

In addition, global trends such as population aging and the rise of chronic diseases are reshaping healthcare needs. Older populations require more complex medical care, often involving multiple comorbidities and chronic disease management. These changes call for a more integrated, patient-centric approach to healthcare, in which Medical Affairs plays a critical role in understanding patient needs, facilitating the development of new therapies, and ensuring their appropriate use.

Increasing Importance of Medical Affairs in Pharmaceutical Business

The shift toward value-based healthcare—where the emphasis is on delivering outcomes rather than simply increasing the volume of treatments—has placed Medical Affairs at the forefront of pharmaceutical company strategy and operations. No longer relegated to a supporting role, Medical Affairs now serves as a key player in shaping research priorities, guiding business decisions, governing operations, and maintaining ethical and regulatory standards, especially in clinical research and product post-approval activities and promotion.

As healthcare becomes more outcomes-driven, Medical Affairs professionals are increasingly tasked with ensuring that a pharmaceutical company's research aligns with patient needs and delivers real value to healthcare systems. These professionals also engage with key external experts (KEEs), HCPs, and patients to gather insights that will inform the development and promotion of new products. Through this engagement, Medical Affairs ensures that products are positioned to be not only commercially successful but also scientifically sound and beneficial for patients.

Medical Affairs also plays a pivotal role in navigating the regulatory landscape. Regulatory bodies worldwide are placing increasing scrutiny on clinical data, requiring robust evidence of efficacy, safety, and cost-effectiveness. Medical Affairs ensures that companies comply with these stringent regulatory requirements.

The evolution of Medical Affairs reflects a broader trend in the pharmaceutical industry toward a more patient-centric, ethical, and scientifically rigorous approach.

This transformation is not simply a reaction to external changes in healthcare and regulation; it is also a proactive move to align with emerging paradigms such as personalized medicine, digital health, and value-based care.

Medical Affairs Mission

Medical Affairs functions as the strategic and operational pillar within pharmaceutical companies, ensuring that clinical research, product development, and healthcare provider interactions are both scientifically credible and aligned with patient needs. Medical Affairs serves as the conscience of pharmaceutical companies, safeguarding scientific integrity and ensuring that patient welfare remains paramount throughout the product lifecycle.

Medical Affairs is not confined to post-market activities. Its role extends across the entire product development process, ensuring that products are scientifically sound and commercially viable. The department also plays an instrumental role in translating clinical research into practical applications, providing healthcare professionals with the data and insights they need to use medicines safely and effectively. By serving as a bridge between R&D, HCPs, and regulatory bodies, Medical Affairs ensures that the voice of science remains integral to pharmaceutical operations.

The mission of Medical Affairs is to enhance patient care by integrating scientific expertise with the development and dissemination of innovative medical solutions. The department is responsible for ensuring that products are developed and used ethically, safely, and effectively. Medical Affairs teams work to bridge the gap between clinical development and the practical application of pharmaceutical products, ensuring that they meet the highest standards of safety and efficacy.

Medical Affairs also plays a critical role in maintaining the scientific credibility of pharmaceutical companies. This involves ensuring that all communications are evidence-based and free from commercial bias, promoting transparency and ethical decision-making throughout the organization. In this way, Medical Affairs helps to build trust with healthcare professionals, regulatory bodies, and patients.

The mission of Medical Affairs is inherently aligned with the broader objectives of pharmaceutical companies, contributing not only to the scientific rigor of medicinal product development but also to the overall business strategy. As pharmaceutical companies increasingly emphasize patient-centered approaches, ethical responsibility, and scientific excellence, Medical Affairs ensures that these goals are embedded in every phase of product development and commercialization (Table 1.1).

Table 1.1 Pharmaceutical company mission, vision, core values, goals, and objectives

What	Definition	Specificity for pharmaceutical company	Example of formulation for innovative pharmaceutical company/Medical Affairs
Mission	A mission defines the core purpose and focus of an organization. It articulates why the company exists and what it aims to achieve in the present.	For a pharmaceutical company, the mission often revolves around improving patient outcomes, advancing medical innovation, and ensuring access to high-quality medicines. It may highlight a commitment to addressing unmet medical needs or specific therapeutic areas.	"To discover, develop, and deliver innovative medicines that improve and extend the lives of patients while ensuring accessibility and affordability worldwide."
Vision	A vision represents the long-term aspiration of an organization, outlining its desired future state or ultimate impact on society.	The vision of a pharmaceutical company typically emphasizes transformative goals, such as eradicating diseases, leading in innovation, or achieving global health equity. It portrays the company's role in shaping the future of healthcare and medical science.	"To be the world's leading innovative pharmaceutical company, driving effective and affordable treatments for the most challenging diseases and ensuring better health and quality of life for patients."
Core values	Core values are the guiding principles and ethical standards that influence a company's culture, behavior, and decision-making processes.	Pharmaceutical companies often prioritize values such as patient-centricity, scientific excellence, integrity, innovation, diversity, and collaboration. These values ensure that the organization adheres to ethical practices and maintains trust with stakeholders.	"Integrity, innovation, and patient-first: We prioritize ethical practices, groundbreaking research, and partnerships to ensure the best outcomes for patients globally."
Goals	Goals are broad, overarching statements that outline what the organization aims to achieve over a defined period.	In the pharmaceutical sector, goals often include developing a certain number of new drugs, achieving market leadership in specific therapeutic areas, or improving operational efficiencies to expedite drug development and delivery.	"To launch three breakthrough therapies in oncology and rare diseases within the next 5 years while ensuring their affordability and accessibility across different regions."
Objectives	Objectives are specific, measurable, and time-bound targets that support the achievement of broader goals.	Pharmaceutical companies focus on objectives such as completing specific clinical trials, securing regulatory approvals within a defined timeline, expanding research pipelines, or achieving sales milestones for particular products.	"To complete phase III clinical trials for two investigational medicinal products to deliver innovative treatments to oncology patients by (specific date), with submission to regulatory authorities by (specific date)."

Medical Affairs Vision

The vision for Medical Affairs extends beyond its historical role as a support function. As the pharmaceutical industry continues to evolve, Medical Affairs is poised to take on an even more critical role in shaping the future of healthcare. By embracing emerging technologies, driving innovation in medical science, and contributing to policy discussions that impact global health outcomes, Medical Affairs is becoming a key influencer in the healthcare landscape.

The future of Medical Affairs will involve greater participation in healthcare policy-making, where the department's insights into medicinal product development, patient care, and clinical outcomes can help shape policies that promote innovation and ethical practices. In addition, Medical Affairs will be at the forefront of advancing personalized medicine, digital health, and real-world evidence initiatives. The adoption of new technologies such as artificial intelligence (AI) and big data analytics will enable Medical Affairs to play a pivotal role in improving patient outcomes and optimizing healthcare delivery.

Medical Affairs will be a driving force behind the integration of digital health initiatives, including the use of AI for medicinal product discovery, the application of big data in understanding patient populations, and the implementation of telemedicine and mobile health solutions. These technologies will not only enhance the ability of Medical Affairs to contribute to medicinal product development but will also improve the department's capacity to engage with healthcare providers, regulators, and patients in meaningful ways.

In this rapidly changing landscape, Medical Affairs will be a key driver of innovation, ensuring that pharmaceutical companies remain at the forefront of scientific discovery while maintaining a commitment to patient-centric healthcare. By contributing to the development and implementation of personalized medicine strategies, Medical Affairs will continue to play a central role in improving patient outcomes and advancing healthcare as a whole.

Medical Affairs Core Values

Upholding high ethical standards is at the core of Medical Affairs. The department is entrusted with ensuring that all scientific and medical information disseminated is unbiased, evidence-based, and free from commercial influence. This ethical responsibility extends to all interactions with healthcare professionals, regulators, and patients, where transparency and integrity are paramount.

Medical Affairs serves as a guardian of scientific integrity within pharmaceutical companies, ensuring that decision-making processes are rooted in sound scientific principles rather than commercial considerations. This role is particularly important in clinical research and medicinal product promotion, where the potential for conflicts of interest can arise. Medical Affairs professionals must maintain a commitment to ethical conduct, ensuring that all activities are conducted with the highest level of integrity and that patient welfare is always prioritized.

Ethical considerations also guide the department's approach to stakeholder engagement. Whether working with KEEs, HCPs, or patient advocacy groups, Medical Affairs must ensure that all interactions are transparent, scientifically accurate, and focused on improving patient care. This commitment to ethics not only enhances the credibility of the department but also strengthens the trust between pharmaceutical companies and the healthcare community.

Navigating the complex and ever-changing regulatory landscape is a central responsibility of Medical Affairs. The department must stay abreast of international regulations, industry codes of practice, and local laws governing pharmaceutical activities. Compliance with these standards is critical to ensuring that pharmaceutical products are approved, marketed, and used in a manner that is safe, ethical, and scientifically justified.

Medical Affairs plays a key role in ensuring that all regulatory requirements are met throughout the product lifecycle. This includes overseeing clinical trial design, ensuring adherence to good clinical practice (GCP), and managing regulatory submissions. The department is also responsible for ensuring that all promotional materials and scientific communications comply with regulatory standards and that any claims made about a product's efficacy and safety are supported by rigorous scientific evidence.

As regulatory frameworks become more stringent, Medical Affairs must also be agile in adapting to new compliance requirements. This includes understanding and implementing new guidelines related to pharmacovigilance, patient safety, and RWE. By ensuring compliance with these evolving regulations, Medical Affairs helps to safeguard the company's reputation and ensure that its products meet the highest standards of safety and efficacy.

Impact of Digital Transformation

Digital transformation is revolutionizing the pharmaceutical industry, and Medical Affairs is at the forefront of this change. The adoption of digital technologies such as artificial intelligence (AI) and big data analytics is transforming how medicinal products are developed, how clinical trials are conducted, and how patient outcomes are measured. For Medical Affairs, this digital revolution provides new opportunities to enhance decision-making processes, improve stakeholder engagement, and drive innovation in patient care.

AI is increasingly being used in medicinal product discovery and development, allowing pharmaceutical companies to identify promising compounds, optimize clinical trial designs, and predict patient responses to therapies. Big data analytics, meanwhile, enable Medical Affairs to analyze RWE and patient-reported outcomes, providing valuable insights into the safety and effectiveness of therapies in diverse patient populations. By leveraging these technologies, Medical Affairs can contribute to more efficient medicinal product development processes and more personalized approaches to patient care.

Digital health solutions such as telemedicine, mobile health apps, and wearable devices are also generating vast amounts of patient data that can be used to monitor treatment outcomes and improve patient engagement. Medical Affairs plays a critical role in integrating these data into clinical decision-making, ensuring that digital health technologies are used effectively to enhance patient care and support the company's broader strategic objectives.

Global health challenges, such as pandemics, antimicrobial resistance, and the rising prevalence of chronic diseases, are shaping the pharmaceutical industry's approach to healthcare. Medical Affairs has a strategic role in addressing these challenges, helping pharmaceutical companies to develop innovative solutions that address pressing global health needs.

During pandemics, for example, Medical Affairs is responsible for ensuring that clinical research is conducted ethically and efficiently, that healthcare providers have access to up-to-date scientific information, and that patients receive safe and effective treatments. In the case of antimicrobial resistance, Medical Affairs can help to promote the responsible use of antibiotics and support the development of new therapies that address resistant infections.

As global health challenges continue to evolve, Medical Affairs will remain a key player in shaping pharmaceutical strategies that prioritize patient safety, public health, and the responsible use of medical resources. By engaging with global health organizations, regulatory bodies, and healthcare providers, Medical Affairs can help to ensure that pharmaceutical companies contribute to the global effort to improve health outcomes and address emerging health threats.

The future of Medical Affairs lies in its ability to embrace digital transformation, lead the development of personalized medicine, and contribute to global health initiatives. By maintaining a commitment to ethical standards, scientific integrity, and patient care, Medical Affairs will continue to be a driving force in the pharmaceutical industry, helping companies to navigate the challenges of an ever-changing healthcare landscape and deliver innovative, patient-centered therapies.

Key Objectives of Medical Affairs

Collaboration with and Strategic Contribution to R&D

Collaboration between Medical Affairs and R&D teams is foundational to the success of pharmaceutical development. The synergy between these two departments ensures that the clinical development process is not only scientifically robust but also aligned with patient needs and current therapeutic landscapes.

Medical Affairs provides vital insights drawn from clinical expertise, patient feedback, and RWE. These insights are essential for the R&D teams to refine their focus, ensuring that they develop treatments with the highest potential to meet unmet medical needs. Through collaboration, R&D teams are equipped with a deeper understanding of the practical challenges faced by healthcare professionals and patients, which allows for more targeted research efforts.

In practice, Medical Affairs helps to guide decisions on which disease areas, patient populations, and unmet clinical needs should be prioritized. Their input into early clinical development ensures that investigational therapies are designed to have a meaningful impact on patient care, improving both clinical trial outcomes and the likelihood of regulatory success.

Medical Affairs can also assist R&D teams to navigate regulatory requirements and emerging trends in healthcare, ensuring that their efforts align with both the needs of the healthcare community and the expectations of regulatory bodies.

The role of Medical Affairs in clinical trial design and execution is pivotal. Medical Affairs professionals offer specialized expertise in clinical methodologies, patient selection, and ethical considerations, which helps to ensure that trials are not only scientifically sound but also clinically relevant. Their involvement often begins at the conceptual stage, helping to define key trial parameters such as inclusion and exclusion criteria, endpoints, and patient-reported outcomes.

In addition to their involvement in trial design, Medical Affairs teams also play a critical role in the interpretation and communication of trial results. Their clinical expertise enables them to translate complex trial data into actionable insights that can inform both clinical practice and future research. By doing so, they ensure that trial outcomes are relevant to HCPs and can be used to improve patient care.

Generation and Dissemination of Scientific Information

One of the core responsibilities of Medical Affairs is the generation and dissemination of scientific information and clinical evidence to key stakeholders, including healthcare professionals, regulatory bodies, and patients. This communication ensures that the latest research findings and therapeutic advancements are shared in a manner that is both scientifically accurate and accessible.

Medical Affairs teams are responsible for translating complex clinical data into formats that are understandable and useful for HCPs, enabling them to make informed decisions about patient care. This role involves the creation of medical literature, clinical guidelines, and other educational materials that are disseminated through scientific conferences, publications, and digital platforms.

Effective communication by Medical Affairs is essential in promoting evidence-based practices in healthcare. By ensuring that clinicians have access to the most up-to-date information, Medical Affairs helps to improve patient outcomes and foster trust between pharmaceutical companies and healthcare providers.

Educational Initiatives for Internal and External Stakeholders

Medical Affairs plays a key role in educating both internal teams and external stakeholders. Within pharmaceutical companies, Medical Affairs professionals ensure that sales and marketing teams are well-versed in the scientific and clinical aspects of the therapies they promote. This ensures that these teams can engage with healthcare professionals in an informed and ethical manner.

Externally, Medical Affairs organizes educational programs for HCPs, offering training on new therapeutic options, clinical guidelines, and disease management strategies. These educational initiatives are designed to improve the clinical knowledge of HCPs, helping them to make better treatment decisions and ultimately improve patient outcomes.

Patient education is another critical area of focus for Medical Affairs. By developing resources that help patients understand their conditions and treatment options, Medical Affairs contributes to more engaged and informed patients who can actively participate in their care. This patient-centric approach is increasingly important in the era of personalized medicine, where treatments are tailored to individual needs and preferences.

Establishing, Developing, and Maintaining Relationships with Key External Experts and Medical Societies

KEEs play a significant role in shaping the healthcare landscape, and the relationships that Medical Affairs maintains with these thought leaders are critical to the success of pharmaceutical products. KEEs are often respected clinicians, researchers, or academic figures who have deep expertise in specific therapeutic areas and can influence both clinical practice and the direction of research.

Medical Affairs is responsible for establishing and nurturing long-term relationships with KEEs, facilitating a two-way communication that benefits both parties. Through these interactions, pharmaceutical companies can gain valuable insights into emerging trends, treatment preferences, and patient needs, while KEEs can stay informed about the latest scientific advancements and therapeutic developments.

The collaboration between Medical Affairs and KEEs is particularly important during product development and launch phases, where KEE insights can help shape clinical strategies, marketing plans, and educational initiatives.

The insights provided by KEEs are instrumental in guiding strategic decisions within pharmaceutical companies. By tapping into the clinical experience and knowledge of these experts, Medical Affairs can ensure that the company's research and commercial strategies are aligned with real-world clinical needs.

KEEs provide unique perspectives on patient care, treatment efficacy, and safety considerations that are invaluable in designing clinical trials, developing new products, and shaping market access strategies. By leveraging these insights, Medical Affairs helps to ensure that pharmaceutical products are positioned effectively and meet the expectations of both HCPs and patients.

In addition to product development, KEEs play a crucial role in post-market surveillance and pharmacovigilance. Their feedback on real-world product performance can help identify potential safety concerns or areas for improvement, enabling Medical Affairs to address these issues proactively and ensure continued product success.

Value Proposition and Market Access Strategies

Medical Affairs plays a crucial role in developing market access strategies, ensuring that pharmaceutical products can enter the market and be adopted by healthcare systems. One of the key responsibilities of Medical Affairs in this regard is to communicate the clinical and economic value of new therapies to payers, health systems, and other stakeholders.

By presenting evidence-based data on a product's efficacy, safety, and cost-effectiveness, Medical Affairs helps to build a strong value proposition that can facilitate reimbursement and access decisions. This is particularly important in markets where healthcare budgets are limited and payers are increasingly focused on value-based care.

The involvement of Medical Affairs in market access strategies also includes the generation of RWE that demonstrates the long-term benefits of therapies in diverse patient populations. By providing robust data on clinical outcomes and economic impacts, Medical Affairs ensures that pharmaceutical products are positioned as valuable solutions to healthcare challenges.

Health Economic Considerations

In today's healthcare environment, the economic impact of treatment options is a critical consideration for both HCPs and payers. Medical Affairs is increasingly involved in health economics and outcomes research (HEOR) to demonstrate the value of pharmaceutical products from both a clinical and economic perspective.

HEOR activities conducted by Medical Affairs include cost-effectiveness analyses, budget impact assessments, and the evaluation of quality-adjusted life years (QALYs) associated with treatment options. These analyses provide payers and HCPs with the data they need to make informed decisions about the allocation of resources and the reimbursement of new therapies.

By integrating health economic considerations into their activities, Medical Affairs ensures that pharmaceutical products are not only clinically effective but also offer value in terms of cost savings and improved patient outcomes. This is particularly important in the context of value-based care, where reimbursement is increasingly tied to the delivery of measurable outcomes rather than the volume of treatments.

Medical Affairs Role in Biosimilars and Novel Therapeutics

Medical Affairs plays a key role in the introduction and market positioning of biosimilars and novel therapeutic agents. With the rise of biosimilars and advancements in biotechnology, Medical Affairs is tasked with ensuring that healthcare professionals understand the scientific principles behind these therapies, including their safety, efficacy, and equivalence to reference products.

In the case of biosimilars, Medical Affairs must navigate complex regulatory environments and address concerns about extrapolation, interchangeability, and immunogenicity. Through educational initiatives and strategic communication, Medical Affairs helps to build confidence in biosimilars and ensure their successful adoption in clinical practice.

For novel therapeutics, particularly those based on personalized medicine and emerging biotechnologies, Medical Affairs is responsible for communicating the clinical relevance of these therapies and ensuring that HCPs understand how to integrate them into patient care. By providing up-to-date scientific information, Medical Affairs helps to position these therapies as valuable treatment options that meet unmet clinical needs.

Ethical Marketing and Compliance

As pharmaceutical companies face increasing scrutiny of their marketing practices, Medical Affairs plays a crucial role in ensuring that promotional activities are conducted ethically and in compliance with regulatory requirements. Medical Affairs teams are responsible for reviewing promotional materials, ensuring that all claims about a product's efficacy and safety are supported by scientific evidence.

Medical Affairs also works closely with compliance teams to ensure that all interactions with healthcare professionals adhere to industry guidelines and legal requirements. This includes ensuring transparency in financial relationships with KEEs, adhering to fair marketing practices, and maintaining the integrity of scientific communications.

By upholding ethical standards in marketing and compliance, Medical Affairs helps to build trust between pharmaceutical companies and HCPs, ensuring that products are promoted responsibly and in a manner that prioritizes patient welfare.

Conclusion

As the healthcare landscape continues to evolve, the importance of Medical Affairs will only grow. The department's ability to collaborate with R&D teams, engage with KEEs, and contribute to market access strategies will be essential in navigating the challenges of a rapidly changing healthcare environment. Medical Affairs plays a central role in ensuring that pharmaceutical products are developed, communicated, and marketed in a manner that is both scientifically rigorous and patient-centered. By maintaining a commitment to scientific integrity, ethical practices, and patient welfare, Medical Affairs will continue to be a driving force in the pharmaceutical industry, shaping the future of healthcare and improving patient outcomes.

Chapter 2
Navigating an Evolving Landscape in the Pharmaceutical Industry

Key Highlights

- Impact of digital transformation (AI, big data, and real-world evidence) on Medical Affairs operations.
- Rising need for regulatory expertise due to evolving compliance requirements.
- Importance of strategic communication and relationship management with healthcare professionals and regulators.
- Role of Medical Affairs in adapting strategies for value-based healthcare and global market expansion.
- Necessity for continuous learning and professional development to keep pace with scientific innovations.

The pharmaceutical industry is experiencing a profound transformation driven by technological innovations, evolving regulatory environments, and shifting healthcare paradigms. Medical Affairs, once primarily a support function, has evolved into a central pillar that bridges clinical development with product commercialization, making it indispensable to pharmaceutical operations. The rapid pace of change across the healthcare sector has forced Medical Affairs to adapt its functions, ensuring its relevance in a dynamic ecosystem that increasingly emphasizes patient outcomes, real-world evidence (RWE), and data-driven decision-making.

Importance of Advancing Medical Affairs Competency and Skill Development

In this evolving landscape, the need for continuous skill and competency development in Medical Affairs is paramount. The transformation within the industry—from traditional pharmaceutical models to those that incorporate advanced technologies and personalized therapies—demands that Medical Affairs

A. Krendyukov, *Medical Affairs' Fundamentals: The Next Generation*,
https://doi.org/10.1007/978-3-031-92588-7_2

professionals possess a diverse and sophisticated skill set (refer Table 2.1). This section explores the key competencies that professionals in this field must develop to remain effective.

- *Adaptation to Technological Innovations*: The integration of cutting-edge technologies, such as artificial intelligence (AI), machine learning, and big data analytics, has significantly impacted how medical information is processed, disseminated, and applied. Medical Affairs professionals must be adept at leveraging these technologies to enhance decision-making processes and enable more personalized and efficient healthcare solutions. These tools are critical for gathering RWE, supporting clinical trials, and optimizing patient care pathways.
- *Advanced Scientific Acumen*: As the complexity of pharmaceutical products increases, particularly in areas such as gene and cell therapies, Medical Affairs professionals are required to possess a deeper understanding of the underlying scientific principles. Keeping pace with advances in fields such as immunotherapy, oncology, and rare diseases is essential for providing accurate scientific guidance and contributing to clinical development strategies.
- *Strategic Communication Skills*: Medical Affairs professionals act as the voice of science within pharmaceutical companies, responsible for translating complex clinical data into information that is actionable for healthcare providers (HCPs), payers, and patients. The ability to convey nuanced scientific findings in a clear and effective manner is critical for fostering trust, driving adoption of new therapies, and enhancing the value of pharmaceutical products in the market.
- *Regulatory and Compliance Expertise*: Given the heightened scrutiny surrounding pharmaceutical marketing and promotion, Medical Affairs teams must be well-versed in the evolving regulatory landscape. A thorough understanding of compliance standards, particularly in emerging markets, is essential for ensuring that all activities meet ethical and legal requirements. This expertise is particularly valuable in mitigating risk and facilitating market access.

Table 2.1 Key differences between knowledge, competencies, skills, behaviors, and attitudes

Aspect	Definition	Key features	Methods of development	Evaluation methods	Example in a professional context
Knowledge	Information, facts, and theoretical understanding acquired through learning or experience	Knowledge is foundational and includes subject-specific insights, principles, and data. In the pharmaceutical context, it might include regulatory requirements, disease biology, or drug development processes	Formal education, reading, research, attending conferences, and self-study	Written tests, certifications, and knowledge assessments	"Understanding the mechanisms of action for targeted oncology therapies"
Competencies	The combination of knowledge, skills, and behaviors required to perform a task effectively in a specific context	Competencies are holistic and context-dependent, involving not only technical proficiency but also interpersonal and organizational skills. They often align with organizational goals and values, such as leadership in cross-functional teams or ethical decision-making	Training, mentorship, practical experience, and leadership development programs	Performance reviews, 360-degree feedback, and peer evaluations	"Demonstrating regulatory compliance competency by successfully leading submissions to the FDA and EMA while managing cross-functional teams"
Skills	The ability to perform specific tasks or activities, often developed through practice and training	Skills are actionable and measurable, encompassing both technical (e.g., data analysis) and soft skills (e.g., communication). Skills are often specific to roles and can evolve with continuous learning and practice	Practice, hands-on activities, role-specific training, simulations, and workshops	Practical demonstrations, simulations, and KPI tracking	"Applying statistical methods to analyze clinical trial data and effectively communicating results in regulatory submissions"

(continued)

Table 2.1 (continued)

Aspect	Definition	Key features	Methods of development	Evaluation methods	Example in a professional context
Behaviors	Actions that reflect organizational values and ethics in professional settings	Behaviors encompass how individuals act and respond in various situations, showcasing their commitment to integrity, collaboration, and patient-centered practices	Observational learning, feedback from colleagues and mentors, and organizational culture reinforcement	Behavioral assessments, feedback surveys, and direct observation	"Maintaining transparency in drug safety communication to regulatory authorities"
Attitudes	A mental outlook that supports growth, effective collaboration, and a willingness to adapt to change	Attitudes influence how individuals approach their roles and responsibilities, fostering a mindset for innovation, ethical practices, and continuous improvement	Self-reflection, workshops on emotional intelligence, and exposure to diverse perspectives	Attitude surveys and feedback from team members and leaders	"Cultivating a patient-first mindset to prioritize health outcomes and innovative solutions"

Impact of Changing Healthcare Dynamics

The changing healthcare landscape presents both challenges and opportunities for Medical Affairs. Shifts in how healthcare is delivered and paid for have altered the role of Medical Affairs in the pharmaceutical industry, necessitating a more dynamic and patient-centered approach.

- *Shift Toward Value-Based Healthcare* (Fig. 2.1): The global move toward value-based care emphasizes outcomes over volume. Medical Affairs now plays a key role in generating evidence that demonstrates the clinical and economic value of pharmaceutical products. This requires robust data to support reimbursement decisions, pricing negotiations, and patient access programs.
- *Patient-Centricity*: In an era where patient-centric models dominate, Medical Affairs must work directly with patients and patient advocacy groups to ensure that patient perspectives are incorporated into every stage of product development. From clinical trial design to patient education programs, understanding patient needs has become a crucial component of the Medical Affairs' mission.
- *Collaboration with External Stakeholders*: The role of Medical Affairs in forging partnerships with academic institutions, HCPs, and research organizations has grown in importance. These collaborations facilitate scientific dialogue, advance clinical research, and foster innovation, ultimately contributing to the development of products that meet unmet medical needs.
- *Globalization and Market Diversification*: As pharmaceutical companies expand into new global markets, Medical Affairs must adapt strategies to fit diverse regulatory, cultural, and healthcare landscapes. A global approach, informed by local insights, ensures that pharmaceutical products are successfully launched and integrated into varying healthcare ecosystems.

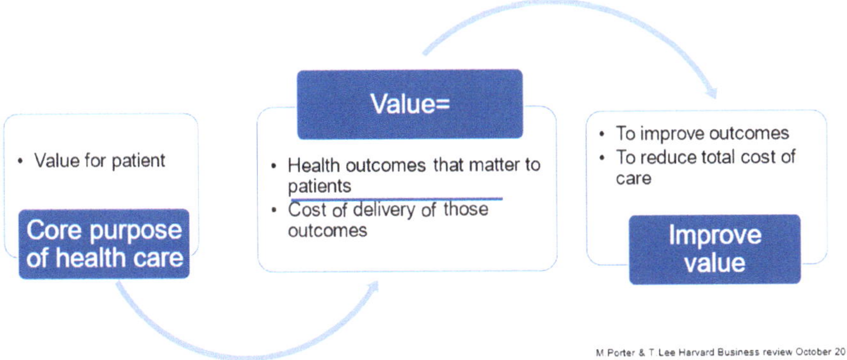

Fig. 2.1 In value-based healthcare, value represents health outcomes that matter to patients and the total costs of delivering those outcomes over the full cycle of care. (This figure was developed by the author for better visualization, based on text, not illustrations, used in the original journal paper M. Porter & T. Lee Harvard Business Review October 2013)

In summary, Medical Affairs sits at the intersection of science and commerce, balancing the need for clinical credibility with the demands of commercialization. The evolving landscape requires professionals to adapt continuously, refining competencies and embracing new challenges to ensure that pharmaceutical companies remain at the forefront of healthcare innovation. The following sections delve into the specific skills, competencies, and strategic approaches that will define the future of Medical Affairs.

Core Competencies and Skills for Modern Medical Affairs

As the role of Medical Affairs evolves, so too must the competencies and skills required for success in the field. Modern Medical Affairs professionals must be versatile, possessing a combination of scientific expertise, communication skills, regulatory knowledge, and technological proficiency. These core competencies enable Medical Affairs teams to effectively navigate the complex pharmaceutical landscape, supporting their organizations' strategic goals and ensuring that products meet both clinical and commercial expectations.

Scientific Acumen and Medical Knowledge

The rapid pace of scientific innovation, particularly in fields such as biotechnology, genomics, and immunotherapy, requires Medical Affairs professionals to engage in continuous learning. Staying current with the latest research, clinical guidelines, and therapeutic advancements is essential for ensuring that Medical Affairs can provide relevant and accurate insights across the product lifecycle.

Ongoing Professional Development: Engaging in continuous education through scientific conferences, peer-reviewed literature, and specialized training programs is critical. Medical Affairs teams must stay abreast of developments in their respective therapeutic areas to provide informed recommendations and contribute meaningfully to medicinal product development strategies.

Expertise in specific therapeutic areas allows Medical Affairs professionals to contribute to strategic decisions during medicinal product development. This specialized knowledge is essential for ensuring that pharmaceutical products address unmet clinical needs and align with the latest scientific discoveries.

Specialized Knowledge: Building deep expertise in therapeutic areas such as oncology, immunology, or rare diseases allows Medical Affairs teams to offer valuable insights during clinical trial design and product development. This expertise positions Medical Affairs as a strategic partner in guiding Research and Development (R&D) efforts and shaping market access strategies.

Communication and Relationship Management

One of the most critical competencies for Medical Affairs professionals is the ability to communicate complex scientific information in a clear and compelling manner. Whether engaging with HCPs, payers, or regulatory bodies, effective communication ensures that all stakeholders understand the value and safety of the pharmaceutical products.

Multifaceted Communication Skills: Medical Affairs professionals must master multiple communication channels, including scientific publications, presentations, digital platforms, and face-to-face interactions. Each medium requires a tailored approach to ensure that the message resonates with the target audience, whether it's explaining the nuances of a clinical trial to a healthcare provider or presenting economic data to a payer.

Establishing and maintaining relationships with Key External Experts (KEEs) is essential for gaining insights into the therapeutic landscape and influencing healthcare practices. These relationships are built on trust and mutual respect, and they enable pharmaceutical companies to stay informed about the latest clinical trends and patient needs.

Strategic Relationship Management: Identifying and engaging KEEs in various therapeutic areas allows Medical Affairs teams to gather valuable feedback on product development, clinical trial design, and market positioning. These relationships also enhance the scientific credibility of pharmaceutical companies, as KEEs often serve as advocates for new therapies and innovations.

Regulatory and Compliance Expertise

Medical Affairs professionals must be proficient in understanding and applying regulatory guidelines across diverse markets. This includes interpreting policies set forth by regulatory bodies such as the United States Food and Drug Administration (FDA), European Medicines Agency (EMA), and local authorities in emerging markets.

Global Regulatory Acumen: Proficiency in regulatory affairs is essential for ensuring that clinical trials, product approvals, and post-market surveillance activities comply with international standards. A deep understanding of the regulatory landscape enables Medical Affairs teams to anticipate challenges and provide guidance on navigating complex regulatory processes.

Compliance with ethical standards and regulatory requirements is critical for maintaining the credibility and integrity of Medical Affairs activities. Ensuring that all interactions with healthcare professionals and patients are transparent and aligned with regulatory guidelines helps to safeguard the reputation of pharmaceutical companies.

Compliance Strategies: Developing and implementing robust compliance frameworks is essential for guiding Medical Affairs activities, particularly in regions with stringent regulations or where interactions with healthcare professionals are subject to scrutiny. This competency ensures that all communications and educational efforts meet legal and ethical standards.

Data Analytics and Real-World Evidence

In the age of big data, Medical Affairs professionals must be proficient in analyzing large datasets to derive actionable insights. Data analytics support decision-making processes by providing evidence on patient outcomes, treatment efficacy, and healthcare trends.

Data-Driven Decision-Making: The ability to leverage advanced data analytics tools enables Medical Affairs teams to make informed strategic decisions, optimize clinical trials, and support RWE generation. This skill is increasingly important as healthcare systems demand more evidence-based approaches to treatment and reimbursement.

RWE is becoming a cornerstone of Medical Affairs strategy, providing insights into the real-world use of pharmaceutical products outside the controlled environment of clinical trials. Incorporating RWE into strategic planning and decision-making processes enhances the relevance of product data and supports lifecycle management.

Real-World Evidence Strategies: Medical Affairs professionals must be adept at integrating RWE into their activities, using patient data from everyday healthcare settings to complement clinical trial data. This evidence is crucial for informing post-market surveillance, optimizing product use, and ensuring continued product access.

Medical Affairs Functions at Different Organizational Levels

The function of Medical Affairs varies significantly across different organizational levels—global, regional, and country-specific—each requiring unique strategies, roles, and responsibilities. While global Medical Affairs is focused on developing overarching medical strategies and standards, regional and country-level Medical Affairs must adapt these strategies to local market dynamics, regulatory requirements, and healthcare ecosystems. Understanding these distinctions and managing coordination across these levels is critical for ensuring the success of Medical Affairs.

Global Medical Affairs

At the global level, Medical Affairs is responsible for providing strategic leadership that aligns R&D, regulatory strategies, and commercialization efforts. Global Medical Affairs teams establish the medical strategies that guide the entire organization, ensuring consistency across regions while also providing flexibility for local adaptation.

Strategic Leadership: Global Medical Affairs teams set the tone for scientific communication, medical education, and engagement with healthcare professionals worldwide. They provide the strategic direction necessary for advancing the company's R&D and commercialization efforts.

Coordinating medical strategies on a global scale requires balancing standardization with localization. While global strategies establish core principles and objectives, these must be adapted to fit the unique regulatory, cultural, and healthcare needs of individual regions and countries.

Standardization and Adaptation: The challenge lies in developing strategies that can be standardized across markets while remaining flexible enough to accommodate local differences. This requires strong coordination between global, regional, and country-level Medical Affairs teams to ensure that global strategies are effectively implemented at the local level.

Regional Medical Affairs

At the regional level, Medical Affairs must address the specific healthcare needs, challenges, and market dynamics of the area. Regional teams tailor global strategies to fit the unique demands of local healthcare systems, patient populations, and market trends.

Customized Regional Strategies: Developing region-specific strategies allows Medical Affairs teams to address healthcare disparities, patient access issues, and regulatory hurdles that may not be present at the global level. This customization ensures that the company's products are positioned appropriately in diverse markets.

Regional Medical Affairs teams must engage directly with local HCPs, payers, and regulatory bodies to ensure the success of pharmaceutical products in the market. Building strong relationships with these stakeholders is critical for understanding local healthcare dynamics and gaining insights into unmet medical needs.

Ecosystem Engagement: Regional Medical Affairs teams must be skilled in navigating the local healthcare ecosystem, including understanding how HCPs make treatment decisions, how payers evaluate cost-effectiveness, and how regulatory bodies assess new therapies.

Country-Level Medical Affairs (see Case Study #1)

Country-level Medical Affairs teams are responsible for implementing global and regional strategies in a way that aligns with local regulatory frameworks, healthcare infrastructure, and cultural considerations. These teams play a crucial role in ensuring that pharmaceutical products meet local needs and comply with country-specific requirements.

Localized Implementation: Tailoring global and regional strategies to fit country-specific needs involves close collaboration with local healthcare professionals, regulators, and payers. This localized approach ensures that the company's products are integrated into local healthcare systems effectively.

Country-specific considerations such as local regulations, cultural nuances, and healthcare access challenges must be carefully managed by Medical Affairs teams. Understanding these factors is essential for ensuring that products are marketed, reimbursed, and adopted successfully in each country.

Cultural and Regulatory Adaptation: Medical Affairs teams must adapt strategies to align with local cultural values, regulatory frameworks, and healthcare practices. This requires a deep understanding of the local market and the ability to navigate its unique challenges.

In brief, Medical Affairs plays a vital role at every organizational level within the pharmaceutical industry. Whether at the global, regional, or country level, Medical Affairs professionals must tailor their strategies to address the unique challenges and opportunities of each market. Effective coordination across these levels is essential for ensuring the success of pharmaceutical products in diverse healthcare environments.

Case Study#1: Transforming Medical Affairs into a Strategic Pillar in Spain

Foundational Steps in Transformation

- *Stakeholder-Centric Framework*: Company AZ's approach to transformation began with a comprehensive assessment of the Spanish healthcare ecosystem. This dual assessment integrated external stakeholder trends with internal organizational insights. The external analysis identified three primary communities influencing healthcare outcomes: patients, clinicians and KEEs, and institutional stakeholders. Each community's specific needs were categorized into archetypes, providing a granular understanding of expectations and gaps.

Key findings included the rising demand for accessible information among patients, enhanced digital integration for clinicians, and sustainable innovation models for institutional stakeholders. Transversal trends, such as the increasing importance of digital tools, collaborative partnerships, and data-driven decision-making, further underscored the need for Medical Affairs to adopt a more dynamic and integrated operational model.

Internal Restructuring and Capability Building: Internally, company AZ established a dedicated Transformation Office in 2020. This multidisciplinary team spearheaded initiatives to align organizational objectives with external trends. Focus areas included upskilling Medical Affairs professionals to navigate digital environments, integrating RWE into decision-making processes, and fostering cross-functional collaboration. These initiatives laid the foundation for creating a cohesive Medical Affairs strategy that aligns with the lifecycle management of pharmaceutical products and enhances stakeholder engagement.

Strategic Pillars for Industry-Wide Adaptation

- While AZ's transformation model is tailored to Spain's decentralized healthcare system, its core principles are universally applicable. Key pillars include:
- *Proactive Stakeholder Engagement*: Medical Affairs departments must engage stakeholders early in the drug development and commercialization process. This involves moving beyond traditional roles to actively contribute to clinical and policy discussions, ensuring alignment with stakeholder priorities from the outset.
- *Integration of Digital and Data Tools*: The rise of digital health technologies necessitates that Medical Affairs teams harness advanced analytics, artificial intelligence, and digital platforms to deliver personalized insights. These tools are pivotal in addressing the growing demand for real-time, evidence-based solutions.
- *Value-Added Services*: As healthcare systems shift toward holistic care models, Medical Affairs must innovate beyond product-focused offerings. This includes developing services that enhance treatment adherence, chronic disease management, and patient education.
- *Global Adaptability with Local Customization*: The healthcare ecosystem varies significantly across regions. For example, while Spain's system emphasizes regional governance, countries like the United States prioritize market access strategies involving private payers. Medical Affairs strategies must therefore balance global consistency with local relevance, addressing unique regulatory, economic, and cultural contexts.

Lessons Learned

- The transformation journey has highlighted several critical lessons for Medical Affairs departments:
- *Early Engagement is Crucial*: Stakeholders increasingly influence pharmaceutical activities from pre-commercialization to post-launch. Establishing early relationships ensures that Medical Affairs' strategic contributions are recognized and valued.
- *Hybrid Interaction Models*: The COVID-19 pandemic accelerated the adoption of hybrid engagement models combining face-to-face and digital interactions. These approaches are essential for overcoming access barriers and fostering meaningful collaborations.
- *Collaboration as a Catalyst for Innovation*: Partnerships with scientific societies, regulatory agencies, and patient advocacy groups are instrumental in co-creating solutions that address systemic healthcare challenges. Initiatives like the specific company AZ-driven program exemplify the collaborative efforts to optimize clinical practice and improve chronic disease management.

Looking ahead, the role of Medical Affairs will continue to expand, driven by the integration of emerging technologies and the growing emphasis on patient-centered care. The journey toward the 2030 vision requires a unified organizational vision that prioritizes shared objectives, streamlined processes, and resource optimization. By fostering a culture of agility and innovation, Medical Affairs can solidify its position as a cornerstone in the pharmaceutical industry.

The transformation of Medical Affairs from a supportive function to a strategic pillar represents a pivotal moment for the pharmaceutical industry. Company AZ's experience offers a valuable blueprint for navigating this transition, demonstrating the importance of stakeholder-centric approaches, digital integration, and adaptive strategies. As Medical Affairs departments worldwide embrace these principles, they are poised to become indispensable partners in shaping the future of healthcare. By aligning their efforts with broader public health objectives, Medical Affairs teams can enhance their impact on patient outcomes, healthcare system efficiency, and industry innovation—ultimately contributing to a healthier, more sustainable world.

Key Differences Between Medical Affairs and Other Functions

In the pharmaceutical industry, the distinction between various functional areas, particularly Medical Affairs, clinical development, and commercial functions, is crucial for the effective operation of each unit (Tables 2.2 and 2.3). While these functions have distinct roles and objectives, they are also interdependent, working together to ensure the successful development, launch, and lifecycle management of pharmaceutical products. This section outlines the key differences and collaborative aspects between Medical Affairs, Clinical Development, and Commercial functions.

Medical Affairs Versus Clinical Development

While Medical Affairs and Clinical Development both play crucial roles in the medicinal product development process, their focus and objectives differ. Clinical Development is primarily concerned with the design, execution, and oversight of clinical trials, ensuring that medicinal products are developed in accordance with regulatory guidelines and scientific rigor. In contrast, Medical Affairs focuses on the medicinal product life cycle, including post-launch activities, including medical education, scientific communication, and gathering RWE.

Role of Medical Affairs: Medical Affairs acts as a bridge between Clinical Development and the healthcare community, ensuring that clinical trial findings are accurately communicated to HCPs and patients. This function ensures that the real-world application of clinical data is understood and that products are used safely and effectively.

Role of Clinical Development: Clinical Development is responsible for overseeing clinical trials, ensuring that they meet regulatory standards and that the data generated is robust enough to support product approval and market access.

Table 2.2 Comparison of Medical Affairs with Clinical Development and Commercial functions

Aspect	Clinical Development	Medical Affairs	Commercial Functions
Primary focus	Design, execution, and oversight of clinical trials to ensure regulatory approval and robustness of the product data	Supporting the medicinal product lifecycle pre- and post-approval & launch, including scientific communication, medical education, and real-world evidence generation	Maximizing market presence and driving product commercial activities through marketing and sales strategies
Objective	Generate high-quality, regulatory-compliant data to achieve product approval & registration	Facilitate the translation of clinical trial data into real-world applications and ensure products are used safely and effectively	Achieve sales growth, market share, and product competitiveness in the defined markets
Stakeholders	Regulatory authorities, clinical investigators, and R&D teams	Healthcare providers, patients, payers, and internal Clinical Development/R&D, Market Access, and Commercial teams	Sales representatives, marketing teams, and external promotional partners
Collaborative efforts	Sharing clinical insights, patient feedback, and safety data as part of clinical development strategies	Ensuring clinical trial outcomes are communicated accurately and effectively to internal and external stakeholders	Aligning product messaging and educational content with scientific data within the intended indication/label to support market readiness
Challenges	Maintaining rigorous adherence to regulatory standards while generating comprehensive trial data	Balancing stakeholder needs and insights with strict compliance and ethical requirements	Aligning marketing and commercial strategies with regulatory and ethical regulation and code of practice
Strategies for alignment	Integrating patient insights and feedback to inform trial design and objectives	Providing scientific guidance to commercial teams to ensure compliant and evidence-based messaging	Collaborating with Medical Affairs to prioritize patient welfare while achieving marketing objectives
Impact on product lifecycle	Early-stage trial design and execution to ensure regulatory success	Bridging clinical insights with pre- and post-approval strategies to enhance real-world applications	Supporting market penetration and adoption through targeted communication and outreach
Examples of success metrics	Number of regulatory approvals and successful Phase II/III (registrational) trials	Number of peer-reviewed publications and medical education programs delivered	Revenue growth, market share increases, and campaign effectiveness

(continued)

Table 2.2 (continued)

Aspect	Clinical Development	Medical Affairs	Commercial Functions
Ethical responsibilities	Ensuring patient safety and compliance with clinical trial standards	Ensuring patient safety and compliance with regulations and applicable standards. Communicating unbiased, scientifically accurate information and maintaining transparency	Aligning marketing strategies with truthful and ethical messaging within the approved indication/label

Medical Affairs intersects with both Clinical Development and Commercial Functions, playing a vital role in ensuring scientific rigor, effective communication, and ethical practices throughout a product's lifecycle. This table provides a comprehensive comparison of their distinct and collaborative roles

Table 2.3 Roles of Clinical Development, Medical Affairs, and Commercial/Market Access teams in key Innovative Medicinal Product (InMP) lifecycle stages

Key communication imperatives and activities	Clinical Development	Medical Affairs and Medical Science Liaison team	Commercial team
InMP life cycle			
Preparation of the launch (pre-launch): educational and awareness	X	✓	X
Launching activities	X	✓	✓
Post-approval/post-launch activities	X	✓	✓
Early engagement with HCPs prior to InMP launch	✓ (Depends on objective)	✓	X
Engagement with HCPs during and after InMP launch	X	✓	✓
InMP clinical data	✓	✓	Depends
Early understanding of standard of care and unmet medical needs	X	✓	Depends
Patient journey	X (depends)	✓	✓
Understanding the clinical evidence gap required for InMP approval and reimbursement	✓	✓	X
Scientific relationship and collaboration with key external experts and key medical societies	Depends	✓	Depends
InMP use in real practice/patients	X	✓	✓
InMP advantages over direct/indirect competitors	X	✓	✓
Promotional campaign according to the InMP label	X	X	✓
Medical information and unsolicited requests for information beyond the InMP label	X	✓	X

Adapted from Krendyukov and Nasy 2020

Collaboration between Medical Affairs and Clinical Development is essential for ensuring that the transition of a medicinal product from clinical trials to market is seamless. This involves sharing clinical insights, safety data, and patient feedback to inform both clinical development and post-market strategies.

Collaborative Efforts: Medical Affairs and Clinical Development must work closely to ensure that clinical development strategies align with market needs and that the findings from clinical trials are effectively communicated to stakeholders. This collaboration is critical for preparing the market for new products and ensuring their successful adoption.

Medical Affairs Versus Commercial Functions

The primary distinction between Medical Affairs and Commercial Functions lies in their objectives. Medical Affairs focuses on scientific communication, ensuring that HCPs, payers, and patients understand the value and proper use of pharmaceutical products. In contrast, Commercial Functions are responsible for maximizing the product's market presence through marketing and sales strategies.

Medical Affairs' Scientific Focus: Medical Affairs ensures that all information provided to stakeholders is scientifically accurate, unbiased, and compliant with regulatory standards. The goal is to promote evidence-based practice and ensure that healthcare providers have the knowledge they need to use products effectively.

Commercial Functions' Marketing Goals: The Commercial team is responsible for developing and executing marketing strategies, driving product awareness, and ensuring market penetration. Their primary focus is on sales growth and market share, which requires a different set of skills and objectives compared to Medical Affairs.

One of the key challenges for Medical Affairs is balancing the diverse needs of stakeholders—such as HCPs, patients, and payers—while adhering to strict compliance and ethical guidelines. In contrast, Commercial Functions must balance the need for aggressive marketing with the imperative to remain compliant with regulations.

Strategies for Balancing Needs: Medical Affairs plays a critical role in ensuring that the company's commercial activities align with scientific evidence and ethical standards. This includes guiding Commercial Teams on how to communicate product benefits in a manner that is compliant with regulatory guidelines and ensures that patient welfare remains the top priority.

In brief, while Medical Affairs, Clinical Development, and Commercial Functions have distinct roles, they are highly interdependent. Their collaboration is essential for ensuring that pharmaceutical products are developed, approved, marketed, and supported in a manner that is scientifically sound, ethically responsible, and commercially successful.

By developing these competencies and understanding the interplay between different functions, Medical Affairs professionals are well-equipped to navigate the evolving pharmaceutical landscape. This section has explored how Medical Affairs continues to shape and respond to industry trends, highlighting its pivotal role in ensuring the success of pharmaceutical products in an increasingly complex and globalized market.

Training and Development Initiatives in Medical Affairs

Continuous professional development for Medical Affairs professionals is no longer optional—it is essential. The industry's rapid evolution, characterized by constant scientific advancements, technological integration, and changing regulations, requires professionals to consistently update their knowledge and refine their skills. As a critical function that bridges clinical research, regulatory affairs, and commercialization, Medical Affairs must stay ahead of these changes to maintain its strategic relevance and effectiveness.

Continuous training and development initiatives enable Medical Affairs professionals to adapt to an evolving landscape. This section focuses on industry-wide competencies, the importance of certifications, specialized training, and the development of both hard and soft skills to underscore the integral role these initiatives play in fostering excellence and career growth within the field. The chapter also discusses the emerging need for digital and patient-centric competencies, which will drive the future of Medical Affairs.

Continuous Professional Development Programs

The pharmaceutical sector, positioned at the cutting edge of medical research and healthcare, necessitates that Medical Affairs professionals continually develop a broad range of competencies. Industry-wide competencies are foundational to success and encompass technical skills such as scientific knowledge and regulatory acumen, as well as soft skills like communication, leadership, and ethics.

Role of Industry-Wide Initiatives: Numerous initiatives led by pharmaceutical associations, educational institutions, and regulatory bodies aim to enhance these competencies. Seminars, workshops, online courses, and conferences are essential platforms for professionals to stay updated on scientific trends, regulatory changes, and the latest therapeutic advancements. These programs, often hosted by industry leaders such as the Medical Affairs Professional Society (MAPS) and the International Society for Pharmacoeconomics and Outcomes Research (ISPOR), focus on both foundational knowledge and emerging trends such as value-based care and health economics.

Impact on Professional Growth: Participation in these initiatives facilitates career advancement by broadening skill sets, expanding professional networks, and keeping professionals informed of the rapidly shifting pharmaceutical landscape.

By staying engaged in continuous learning, Medical Affairs professionals are better equipped to lead discussions on new therapies, collaborate with R&D teams, and communicate complex data to HCPs and payers.

In addition to broader initiatives, certifications and specialized training programs play a crucial role in validating and honing the skills of Medical Affairs professionals. These targeted learning opportunities provide a deeper dive into specific competencies and often serve as benchmarks for professional expertise.

Potential certifications, such as the Medical Affairs Specialist (MAS) and Medical Science Liaison (MSL) credentials, are widely recognized as indicators of professional competence in the field. These certifications validate a professional's ability to navigate the complexities of Medical Affairs and are often prerequisites for leadership positions within the pharmaceutical industry. The MAS/MSL certification, for example, covers critical areas like compliance, clinical trials, and strategic communication, ensuring that certified professionals meet industry standards.

Specialized Training Programs: Specialized programs offer hands-on experience and in-depth knowledge in niche areas of Medical Affairs such as regulatory affairs, pharmacovigilance, and clinical trial design. Professionals who pursue specialized training in data analytics, for example, can enhance their ability to interpret RWE, which is increasingly critical for market access strategies and post-launch studies.

In summary, continuous professional development through industry-wide programs, certifications, and specialized training is indispensable for Medical Affairs professionals. These initiatives ensure that they remain proficient in the latest knowledge, adapt to evolving industry demands, and uphold the highest standards of excellence in their roles.

Emerging Areas of Training

The integration of digital tools and emerging technologies into pharmaceutical operations is reshaping how Medical Affairs operates. The use of AI, machine learning, big data analytics, and digital communication platforms has become integral to the role, making digital literacy a critical competency.

Need for Digital Integration: AI and big data analytics, for example, enable Medical Affairs teams to efficiently process large datasets and extract actionable insights. These insights inform medicinal product development, clinical trial design, and post-market surveillance. Training initiatives that focus on these technologies, such as programs on the use of AI for predicting patient outcomes or analyzing RWE, are becoming increasingly essential. Medical Affairs professionals who understand digital health applications can better engage with stakeholders in a rapidly digitalizing healthcare environment.

Impact on Medical Strategies: Digital tools also play a role in optimizing communication with healthcare professionals and patients. Multi-channel communication strategies, including webinars, digital conferences, and social media engagement, are essential for disseminating medical information widely and efficiently. As the digital transformation of the pharmaceutical industry continues, Medical Affairs professionals must be proficient in using these platforms to enhance their outreach and educational efforts.

While technical expertise remains a cornerstone of Medical Affairs, soft skills are equally vital for success. As Medical Affairs teams frequently interact with a diverse range of stakeholders, from R&D departments to external HCPs, leadership, communication, and strategic thinking become key competencies.

Importance of Leadership and Communication Skills: Effective leadership within Medical Affairs teams ensures that scientific data is not only accurate but also actionable. Leaders in Medical Affairs must facilitate cross-functional collaboration, manage teams, and ensure that strategic goals align with the broader objectives of the pharmaceutical company. Communication, meanwhile, is crucial in conveying complex clinical data to non-specialist audiences, ensuring that all stakeholders—whether they are healthcare professionals, regulatory bodies, or patients—fully understand the value and safety of the pharmaceutical products.

Strategic Thinking: As Medical Affairs becomes more involved in guiding the strategic direction of pharmaceutical companies, the ability to think critically and strategically is increasingly important. Training programs that focus on decision-making, problem-solving, and strategic planning equip professionals to better navigate the evolving pharmaceutical landscape and contribute to long-term organizational success.

With pharmaceutical companies expanding into global markets, Medical Affairs professionals need to be culturally competent and aware of the diverse healthcare systems, regulatory frameworks, and market dynamics of different regions.

Global Market Training: Training programs that focus on global healthcare trends, regulatory policies, and cross-cultural communication are invaluable for professionals working in international markets. These programs prepare professionals to navigate regulatory complexities, address regional healthcare challenges, and engage with HCPs and regulators across different cultural and geographical contexts.

In an industry as highly regulated as pharmaceuticals, maintaining ethical standards and compliance is non-negotiable. Medical Affairs professionals must be trained to navigate an increasingly complex regulatory landscape, ensuring that all activities—whether clinical research, product promotion, or stakeholder engagement—adhere to legal and ethical standards.

Training in compliance includes understanding the laws and guidelines that govern interactions with healthcare professionals, patients, and regulators. This training helps professionals navigate complex areas such as off-label promotion, transparency requirements, and the ethical dissemination of medical information. The Pharmaceutical Research and Manufacturers of America (PhRMA) Code on Interactions with Healthcare Professionals, for example, provides a framework that guides ethical decision-making within the industry.

Incorporating case-based learning into ethics and compliance training allows professionals to apply theoretical knowledge to real-world scenarios, enhancing their ability to navigate complex regulatory environments. These case studies help professionals anticipate and mitigate compliance risks, ensuring that pharmaceutical companies maintain their credibility and trustworthiness in the eyes of HCPs and regulators.

Future Trends and Emerging Competencies in Medical Affairs

As the pharmaceutical industry continues to evolve, the role of Medical Affairs is expanding in new and transformative ways. Driven by technological advancements, a growing emphasis on patient-centric care, and global healthcare trends, the competencies required for success in Medical Affairs are also evolving. This section examines the emerging trends and future skills that will define the next generation of Medical Affairs professionals, including digital literacy, patient engagement, and the growing importance of Health Economics and Outcomes Research.

Technological Advancements and Digital Literacy & Acumen

In an increasingly digital world, Medical Affairs professionals must integrate cutting-edge technologies into their day-to-day operations. AI, machine learning, and big data analytics are reshaping how clinical data is analyzed, shared, and used to inform decisions.

Need for Digital Integration: Digital tools allow Medical Affairs professionals to make more informed, data-driven decisions. For example, AI can help in predicting patient outcomes, identifying adverse event patterns, or optimizing clinical trial recruitment. These tools not only enhance efficiency but also provide deeper insights into healthcare trends and treatment efficacy.

Impact on Medical Strategies: As digital tools become more integral to pharmaceutical strategies, Medical Affairs must develop competencies in utilizing digital health solutions. This includes proficiency in using electronic health records (EHRs), telemedicine platforms, and mobile health applications to gather RWE and engage with patients and HCPs.

Digital literacy is becoming a core competency in Medical Affairs, going beyond basic technical proficiency to encompass a deep understanding of how digital tools can be strategically applied to advance medical and business goals.

Building Digital Competency: Medical Affairs professionals must be proficient in using digital communication platforms, analyzing digital health data, and leveraging AI-powered tools to enhance patient care and product development. Training programs that focus on digital literacy enable professionals to stay ahead of technological trends and utilize these tools effectively in their roles.

Patient-Centric Competencies

Patient-centricity is becoming a central focus in the pharmaceutical industry, with Medical Affairs playing a key role in incorporating patient insights into clinical development and post-market strategies.

Understanding Patient Needs: The ability to gather and interpret patient feedback, whether through patient advocacy groups, direct engagement, or digital health platforms, is a critical competency. Medical Affairs professionals must learn how to translate these insights into actionable strategies that inform medicinal product development, clinical trial design, and patient education programs.

Patient engagement involves actively involving patients in the decision-making process regarding their healthcare. For Medical Affairs, this means creating platforms for meaningful dialogue with patients and ensuring that their perspectives are reflected in treatment strategies.

Skills for Patient Engagement: Developing competencies in patient communication, advocacy, and education ensures that Medical Affairs professionals can facilitate more informed patient participation in their care. Training in empathy, communication, and cultural sensitivity is essential for effective patient engagement, especially in diverse global markets.

Health Economics and Outcomes Research (HEOR)

As value-based healthcare becomes more prevalent, Medical Affairs must develop competencies in HEOR to demonstrate the clinical and economic value of pharmaceutical products.

HEOR Competencies: Skills in cost-effectiveness analysis, budget impact modelling, and patient-reported outcomes are critical for Medical Affairs professionals. These competencies allow Medical Affairs teams to engage effectively with payers, regulators, and HCPs to ensure that pharmaceutical products are reimbursed and adopted based on robust economic evidence.

Implications for Pharmaceutical Companies

The changing landscape of Medical Affairs has significant implications for pharmaceutical companies, particularly in how they recruit, retain, and develop talent; structure their departments; and foster cross-functional collaboration. To remain competitive, companies must adopt strategic approaches that align with the evolving competencies and expectations of Medical Affairs.

Strategic Recruitment and Talent Management

As the competencies required in Medical Affairs evolve, pharmaceutical companies must align their recruitment strategies to attract professionals with the necessary skills in digital literacy, patient engagement, and HEOR.

Competency-Based Recruitment: Implementing a competency-based approach ensures that companies hire individuals whose skills align with the future needs of Medical Affairs. For example, companies may prioritize candidates with expertise in data analytics, digital health, or global regulatory affairs.

Retaining top talent and planning for future leadership is essential for ensuring continuity and growth within Medical Affairs.

Companies must create environments that support professional growth, such as offering opportunities for continuous learning, career advancement, and leadership development. Succession planning also ensures that companies are prepared for future leadership transitions, minimizing disruption and ensuring organizational stability.

Adapting Organizational Structures

Pharmaceutical companies must adapt their organizational structures to incorporate new roles and competencies, such as digital health specialists, patient engagement teams, and HEOR experts.

Organizational restructuring may involve creating new teams focused on data analytics, patient outcomes, or digital health, ensuring that Medical Affairs can respond to the industry's evolving demands.

Collaboration between Medical Affairs, R&D, Regulatory Affairs, and Commercial teams is essential for creating a cohesive organizational strategy.

Cross-functional collaboration can be enhanced through structured meetings, shared goals, and integrated workflows, ensuring that all teams are aligned in their efforts to develop and commercialize pharmaceutical products.

Conclusion

The future of Medical Affairs lies in its ability to adapt to new trends and technologies while also developing competencies in patient engagement, HEOR, and digital literacy. Training and development initiatives are critical for ensuring that Medical Affairs professionals remain at the forefront of these changes, positioning themselves as strategic leaders within the pharmaceutical industry. By aligning recruitment, retention, and organizational strategies with these evolving competencies, pharmaceutical companies can foster a more agile, innovative, and patient-centered approach to healthcare.

Chapter 3
Medical Affairs Key Activities

Key Highlights

- Medical Affairs' role in scientific communication, ensuring accurate and compliant dissemination of medical information.
- Engagement with KEEs and medical societies to influence clinical practice and treatment adoption.
- Integration of real-world evidence and digital health solutions to improve patient outcomes.
- Ethical and regulatory compliance as foundational pillars of Medical Affairs operations.
- Medical Affairs' contribution to value-based healthcare through health economics and market access strategies.

In the dynamic and multifaceted pharmaceutical industry, Medical Affairs has evolved into a strategic linchpin, playing a pivotal role in aligning clinical development with commercialization efforts. Medical Affairs bridges scientific research, regulatory compliance, and market strategy, ensuring that pharmaceutical products are not only safe and effective but also meet the needs of patients, healthcare providers (HCPs), and regulatory authorities.

Evolution and Significance of Medical Affairs in the Pharmaceutical Industry

The responsibilities of Medical Affairs have expanded dramatically over time and now engage with every phase of the product lifecycle, from the early stages of medicinal product development to post-market surveillance. This reflects the industry's broader shift toward evidence-based, patient-centric models of healthcare, which prioritize clinical efficacy, safety, and real-world outcomes.

A. Krendyukov, *Medical Affairs' Fundamentals: The Next Generation*,
https://doi.org/10.1007/978-3-031-92588-7_3

Expansion of Responsibilities: Medical Affairs' roles now include conducting health economics analyses, engaging with patient advocacy groups, and managing regulatory compliance issues. These expanded responsibilities demand a unique blend of scientific rigor, business acumen, and strategic insight, which enables Medical Affairs to act as a key driver of product success.

The ability of Medical Affairs to interpret clinical data, gather real-world evidence (RWE), and integrate stakeholder feedback makes it crucial for guiding corporate strategy. This integration helps ensure that pharmaceutical products are aligned with both market needs and regulatory expectations, contributing to overall commercial success and long-term viability.

Linking Medical Affairs Activities to Overall Organizational Goals

Medical Affairs plays an essential role in aligning its activities with the broader goals of pharmaceutical companies. This alignment ensures that the scientific and clinical expertise within Medical Affairs directly informs corporate strategy and operational decisions.

Alignment with Organizational Strategy: From clinical trial design to post-market surveillance, Medical Affairs' activities must be tightly coordinated with broader organizational objectives. For example, its involvement in market access strategies ensures that products are not only clinically effective but also economically viable and accessible to patients.

Driving Commercial Success: By providing insights into market needs, regulatory landscapes, and clinical data, Medical Affairs drives the commercial success of pharmaceutical products. The department's ability to generate trust among HCPs and patients is critical for product adoption and sustained market growth.

Overview of Key Medical Affairs Activities in the Pharmaceutical Industry (Table 3.1)

Medical Affairs activities are fundamental to the success of pharmaceutical companies, ensuring that products are developed, marketed, and utilized in compliance with both scientific and ethical standards. This section provides an in-depth overview of key activities within Medical Affairs, including scientific communication, KEE engagement, medical education, and the organization of advisory boards. These activities collectively serve to bridge the gap between clinical development and commercialization, supporting the strategic objectives of pharmaceutical companies.

Table 3.1 Medical Affairs (MA): objectives, key activities, and key competencies

MA key objectives	MA key activities	MA key deliverables
Innovate evidence and insight generation	Scientific communication (congresses, publications, etc.)	Create a medical strategy for the medicinal products and make sure to clearly deliver it to R&D, Commercial, and other internal stakeholders
Accelerate the access to the treatment	Medical, clinical, and scientific training and education	Developing clinical research concepts, study synopses, and protocols for investigational drugs or currently approved medicinal products in new indications
Transform and personalize medical engagement	Later phase (IIIb-IV) clinical studies, RWE and evidence/data generation and utilization, post-approval commitments, Investigator Initiated Trials, research grants, market access and health economic outcomes research, etc.	Develop, manage, and support a Medical Science Liaison group
Internal strategic leadership and medical direction	Engagement and collaboration with key external experts, medical societies, patients, and patient advocacy groups/ organizations	Develop, implement, and support product safety and pharmacovigilance capabilities
	Advisory Boards and Expert Panels	Develop and implement medical and scientific communication and disease (therapeutic area) awareness strategies and plans
	Medical Affairs strategy development and implementation	Develop and execute product-specific life cycle Medical Affairs strategy and plans
	Enabling and supporting activities for pharmacovigilance, non-promotional review, internal approval for commercial materials, medical information, etc.	Developing and supporting health economic outcomes research and RWE for long-term data and return on investment for new molecules
	Compliance and risk management	Develop, implement, and support a medical information platform to deal with unsolicited requests from HCPs
		Develop and implement key external experts, medical society, and patient group engagements and activity plans

Scientific Communication and Publications

One of the core responsibilities of Medical Affairs is to ensure the accurate and timely dissemination of scientific information. This involves communicating complex clinical and scientific data to internal stakeholders (such as Research &

Development [R&D], Regulatory, and Commercial teams) and external stakeholders (including HCPs, regulatory bodies, and patient advocacy groups).

Internal Communication: Internally, Medical Affairs ensures that teams across the company have access to the most current scientific knowledge. This facilitates informed decision-making throughout the product lifecycle, from R&D strategy to post-market surveillance.

External Communication: Externally, Medical Affairs serves as a conduit for transmitting critical information to HCPs and regulators. This communication ensures that the safety, efficacy, and real-world applicability of pharmaceutical products are well understood by those who prescribe, approve, or use them.

Medical Affairs is responsible for the strategic planning and execution of medical publications, which include peer-reviewed journal articles, clinical trial results, and scientific white papers. This function is vital for establishing the scientific credibility of a pharmaceutical product.

Strategic Publication Planning: Publication strategies must be carefully planned to ensure that key data is disseminated to the right audiences in a timely and ethical manner. Medical Affairs teams oversee the publication process, from manuscript development to submission, ensuring that scientific findings are transparent and accessible to the medical community.

Key External Experts (KEE) and Medical Societies Engagement

Engaging with KEEs and the medical community is one of the most impactful activities within Medical Affairs. KEEs are influential healthcare professionals and researchers who provide valuable insights into clinical practice and patient care, guiding the strategic direction of pharmaceutical companies.

KEE Identification and Engagement: Medical Affairs teams are responsible for identifying KEEs in various therapeutic areas and establishing relationships with them. These relationships are based on mutual respect and the exchange of scientific knowledge, which can influence the development, launch, and adoption of new therapies.

KEEs provide critical insights into unmet medical needs, therapeutic trends, and patient outcomes, all of which are invaluable for shaping the strategic direction of pharmaceutical companies.

Strategic Value of KEE Insights: By leveraging KEE insights, Medical Affairs can influence clinical trial design, product development, and market access strategies. For example, KEE feedback may highlight specific patient populations that would benefit most from a new therapy, leading to more targeted and effective clinical trials.

Medical and Clinical Education and Training

Continuous learning is crucial for the effectiveness of pharmaceutical professionals, given the rapid pace of scientific advancements and evolving regulatory requirements. Internal training programs, developed and conducted by Medical Affairs, help ensure that Medical Science Liaisons and cross-functional teams (e.g., Commercial) remain at the forefront of their fields, equipped with the latest knowledge and skills.

Development of Training Programs: Training programs for pharmaceutical professionals cover a wide range of topics, from scientific and clinical knowledge to regulatory compliance and health economics. These programs are designed to enhance both technical and soft skills, ensuring that teams can effectively engage with both internal and external stakeholders.

In addition to internal training, Medical Affairs is responsible for educating healthcare professionals and other external stakeholders on the latest clinical data, therapeutic guidelines, and best practices.

External Educational Programs: These programs are designed to keep healthcare professionals informed about new treatments and clinical data, ensuring that they have the knowledge necessary to make informed treatment decisions. This educational role is critical for building trust and ensuring that new therapies are appropriately integrated into clinical practice.

Advisory Boards and Expert Panels

Medical Affairs organizes and facilitates advisory boards and expert panels to gather insights from healthcare professionals and researchers. These forums provide a platform for discussing therapeutic advancements, clinical trial designs, and post-marketing strategies.

Advisory Board Management: Medical Affairs teams are responsible for convening these meetings, selecting participants, and ensuring that discussions are productive and aligned with the company's strategic goals.

The insights gained from advisory boards and expert panels are critical for shaping Medical Affairs strategies. By integrating expert opinions into clinical development and market access strategies, Medical Affairs can ensure that products meet both clinical and market needs.

Leveraging Expertise: Expert opinions help inform key decisions related to clinical trial design, regulatory submissions, and patient engagement strategies. This collaboration ensures that pharmaceutical products are scientifically validated and clinically relevant.

Real-World Evidence (RWE) and Evidence/Data Generation and Utilization

The collection and use of RWE are becoming increasingly important for Medical Affairs. RWE provides insights into the effectiveness, safety, and patient outcomes of medicinal products in real-world settings, complementing the data gathered from controlled clinical trials.

Real-World Evidence Utilization: Medical Affairs teams are tasked with gathering and analyzing RWE to inform post-marketing strategies, support regulatory submissions, and provide healthcare professionals with a broader understanding of how a product performs outside of clinical trials.

Digital Health and Telemedicine

As the healthcare landscape becomes more digitized, Medical Affairs is increasingly involved in leveraging digital health solutions and telemedicine platforms to enhance patient engagement and education.

Integration of Digital Health Strategies: Medical Affairs can use digital platforms to disseminate medical information, monitor patient outcomes, and engage with healthcare professionals in virtual settings. This digital transformation is reshaping how Medical Affairs interacts with stakeholders, providing new opportunities for education, communication, and data collection.

Patient Advocacy and Engagement with Patient Groups

Patient advocacy is gaining prominence within Medical Affairs as the pharmaceutical industry adopts a more patient-centric approach to medicinal product development. Engaging with patient advocacy groups and incorporating patient feedback into medical strategies are critical for developing therapies that truly meet patient needs.

Patient Advocacy Programs: By developing and maintaining patient advocacy programs, Medical Affairs can gather valuable insights into patient experiences and preferences, which in turn informs medicinal product development and marketing strategies.

Compliance and Risk Management

Compliance with regulatory standards and ethical practices is fundamental to the success and credibility of Medical Affairs activities. A robust compliance framework ensures that Medical Affairs can operate within legal and ethical boundaries while maximizing its strategic impact.

Risk Management and Compliance: Medical Affairs teams must implement comprehensive risk management protocols to navigate the complex regulatory landscape and mitigate any potential compliance risks associated with their activities.

Market Access and Health Economics

As pharmaceutical companies increasingly focus on value-based healthcare, Medical Affairs plays a key role in demonstrating the economic value of new therapies.

Market Access and Health Economics: By integrating health economics and market access strategies into Medical Affairs activities, pharmaceutical companies can better demonstrate the cost-effectiveness of their products to payers and regulators, facilitating reimbursement and market adoption.

In summary, Medical Affairs activities are essential for ensuring that pharmaceutical products are developed, marketed, and utilized in ways that meet the highest standards of scientific integrity, regulatory compliance, and patient care. The role of Medical Affairs has expanded significantly in recent years, encompassing a diverse range of responsibilities, from scientific communication to patient advocacy and market access. By continuing to adapt to emerging trends—such as digital transformation, RWE, and patient-centric healthcare—Medical Affairs will remain a key driver of innovation and success within the pharmaceutical industry.

Cross-Functional Collaboration

Cross-functional collaboration is now an integral part of Medical Affairs' role, facilitating the integration of scientific, clinical, commercial, and regulatory strategies. This section explores the collaborative activities of Medical Affairs, focusing on its interactions with Clinical Development, Commercial teams, and regulatory bodies, all of which contribute to the success of pharmaceutical products.

Medical Affairs acts as a bridge between these departments, ensuring a cohesive and streamlined approach to medicinal product development and commercialization. The collaborative nature of this role is not only essential for achieving organizational goals but also for fostering innovation, maintaining regulatory compliance, and supporting ethical practices. Through effective cross-functional collaboration, Medical Affairs enhances the value proposition of pharmaceutical products and ensures that they meet the needs of HCPs, patients, and other stakeholders.

Collaboration with Clinical Development

The collaboration between Medical Affairs and Clinical Development is vital for transitioning a medicinal product from clinical trials to the marketplace. Medical Affairs provides critical insights into the clinical data generated during trials, ensuring that it is communicated effectively to stakeholders, including healthcare professionals, payers, and patients. This collaboration facilitates the integration of clinical trial results into broader medical strategies, aligning product development with patient needs and regulatory requirements.

Integration of Clinical Insights: Medical Affairs professionals work closely with Clinical Development teams to translate complex clinical data into actionable insights. This requires a deep understanding of clinical trial design, endpoints, and outcomes. By collaborating with clinical teams, Medical Affairs ensures that clinical insights are aligned with real-world medical practices, guiding the strategic direction of the product.

This collaboration ensures that clinical trial data are not only scientifically sound but also meaningful to HCPs and patients. This alignment is essential for designing clinical trials that meet regulatory standards while addressing unmet medical needs.

Seamless Information Exchange for Integrated Product Development Strategies

Effective information exchange between Medical Affairs and Clinical Development teams is critical for developing integrated medicinal product development strategies. This collaboration involves the seamless transfer of data, insights, and feedback between the two departments, ensuring that clinical trial designs are optimized and that trial outcomes are effectively communicated to stakeholders.

Facilitating Information Flow: Medical Affairs plays a key role in establishing communication channels and collaborative platforms that enable the smooth flow of information between departments. By fostering open communication, Medical Affairs helps ensure that Clinical Development strategies are aligned with medical and commercial objectives, leading to more effective product development and launch strategies.

By providing clinical teams with real-world data and market insights, Medical Affairs enhances the design and execution of clinical trials, ensuring that they produce relevant, actionable results. This collaborative approach is essential for developing medicinal products that are both clinically effective and commercially viable.

Collaboration with Commercial Functions

Medical Affairs is increasingly involved in shaping the commercial strategies of pharmaceutical products, ensuring that marketing and sales initiatives are scientifically accurate and aligned with clinical data. This collaboration is essential for bridging the gap between scientific communication and commercial objectives, ensuring that pharmaceutical products are marketed ethically and effectively.

Collaborative Marketing Strategies: Medical Affairs works closely with marketing teams to develop strategies that reflect the scientific and therapeutic value of pharmaceutical products. This collaboration ensures that promotional materials, product messaging, and sales strategies are grounded in evidence-based medicine and are compliant with regulatory standards.

By aligning with Commercial Functions, Medical Affairs helps ensure that product messaging is consistent across all platforms, from scientific publications to promotional materials. This alignment builds trust with HCPs and payers, fostering long-term relationships that support product adoption and market success.

One of the primary challenges in cross-functional collaboration between Medical Affairs and Commercial Functions is balancing scientific rigor with commercial goals. While marketing teams focus on promoting the benefits of a product, Medical Affairs is responsible for ensuring that all claims are backed by robust clinical data and comply with regulatory requirements.

Ethical Commercial Communication: Medical Affairs plays a key role in reviewing and approving marketing materials, ensuring that they are scientifically accurate, ethically sound, and compliant with industry regulations. This oversight is critical for maintaining the integrity of the company and building trust with healthcare providers and regulatory bodies.

By maintaining this balance, Medical Affairs ensures that Commercial teams can promote products effectively without compromising on scientific integrity or ethical standards. This collaborative approach helps pharmaceutical companies navigate the complex regulatory environment while achieving commercial success.

Regulatory Interactions and Collaborations

Collaboration between Medical Affairs and Regulatory departments is essential for ensuring that pharmaceutical products meet the stringent requirements set by regulatory agencies such as the U.S. Food and Drug Administration (FDA) and the European Medicines Agency (EMA). Medical Affairs works closely with regulatory teams to ensure that all aspects of medicinal product development, from clinical trials to marketing, comply with regulatory guidelines.

Collaborative Compliance Strategies: Medical Affairs and Regulatory teams collaborate to develop strategies that ensure compliance with evolving regulatory standards. This includes submitting clinical trial data for approval, preparing regulatory filings, and ensuring that promotional materials meet all legal requirements. Medical Affairs provides the scientific expertise needed to navigate complex regulatory processes, ensuring that products are approved and marketed in compliance with local and global regulations.

By fostering strong relationships with Regulatory teams, Medical Affairs helps pharmaceutical companies anticipate and address compliance issues, reducing the risk of regulatory delays and ensuring that products reach the market in a timely manner.

Engaging proactively with regulatory agencies is a key function of Medical Affairs. By initiating early and ongoing discussions with regulators, Medical Affairs can address potential challenges and ensure that products are developed and approved in accordance with regulatory expectations.

Regulatory Agency Engagement: Medical Affairs teams often lead the engagement with regulatory agencies, providing clinical data and scientific justifications to support medicinal product approval processes. This proactive engagement helps identify potential regulatory hurdles early in the development process, allowing companies to adjust their strategies and avoid costly delays.

Integration of Technology and Innovation in Collaboration

Technological innovations are reshaping how Medical Affairs collaborates with other departments. Digital tools, data analytics platforms, and virtual collaboration technologies are becoming increasingly important for enhancing communication, data sharing, and project management across functions.

Embracing Technological Advances: Medical Affairs teams are increasingly using digital platforms to facilitate cross-functional collaboration. Tools such as cloud-based data sharing, virtual meetings, and project management software enable real-time collaboration between Clinical, Commercial, and Regulatory departments, making it easier to coordinate efforts across global teams.

By embracing technology, Medical Affairs can streamline its operations and improve the efficiency of its collaborative efforts, ultimately leading to more effective medicinal product development and commercialization strategies.

Focus on Collaborative Decision-Making Processes

Collaborative decision-making processes are essential for aligning the strategic objectives of Medical Affairs with those of other departments. By involving multiple stakeholders in key decisions, pharmaceutical companies can ensure that their strategies are informed by a diverse range of perspectives, leading to more robust and effective outcomes.

Enhanced Decision-Making Frameworks: Medical Affairs plays a key role in facilitating collaborative decision-making processes, bringing together representatives from Clinical, Commercial, and Regulatory teams to discuss strategic objectives, assess risks, and make informed decisions. These discussions ensure that all perspectives are considered, leading to more comprehensive and aligned decision-making.

This collaborative approach helps pharmaceutical companies respond more quickly to market changes, regulatory updates, and new scientific data, ensuring that their strategies remain relevant and effective.

Emphasis on Collaborative Training and Development

Joint training initiatives between Medical Affairs and other functional teams can enhance collaboration by fostering a better understanding of each team's roles, responsibilities, and challenges. These programs provide an opportunity for teams to develop shared knowledge and skills, strengthening cross-functional relationships and improving overall collaboration.

Joint Training Initiatives: Medical Affairs can collaborate with Clinical, Commercial, and Regulatory teams to develop joint training programs that cover key topics such as regulatory compliance, clinical trial design, and market access strategies. These initiatives help build a shared understanding of the medicinal product development process, improving communication and collaboration between departments.

By investing in collaborative training programs, pharmaceutical companies can foster a more cohesive and integrated workforce, leading to more effective cross-functional collaboration and better overall outcomes.

Consideration of Ethical Aspects in Cross-Functional Activities

Ethical considerations are central to the success of cross-functional collaboration in Medical Affairs. As teams work together to develop and commercialize pharmaceutical products, they must ensure that their activities are conducted in an ethical manner, particularly when commercial, clinical, and medical interests intersect.

Ethical Considerations in Collaboration: Medical Affairs plays a key role in ensuring that all collaborative efforts uphold the highest ethical standards. This includes ensuring that clinical data is reported accurately, that promotional materials are truthful and compliant with regulatory guidelines, and that patient safety and well-being are prioritized in all decision-making processes.

By maintaining a strong focus on ethics, Medical Affairs ensures that cross-functional collaborations contribute to the development of safe, effective, and ethically sound pharmaceutical products.

Integration of Patient-Centric Approaches

The growing emphasis on patient-centric healthcare is reshaping how pharmaceutical companies approach medicinal product development and commercialization. Medical Affairs plays a key role in integrating patient perspectives into cross-functional collaboration efforts, ensuring that the needs and preferences of patients are considered at every stage of the product lifecycle.

Patient-Centric Collaboration: Medical Affairs can work with Clinical, Commercial, and Regulatory teams to develop patient-centric strategies that prioritize patient needs and outcomes. This includes engaging with patient advocacy groups, incorporating patient feedback into clinical trial designs, and ensuring that marketing and communication efforts address patient concerns.

By fostering patient-centric collaboration, Medical Affairs helps ensure that pharmaceutical products are not only clinically effective but also aligned with the needs and expectations of patients and HCPs.

Conclusion

Cross-functional collaboration is a cornerstone of Medical Affairs' role in the pharmaceutical industry. By working closely with Clinical Development, Commercial teams, and regulatory bodies, Medical Affairs ensures that pharmaceutical products are developed, marketed, and approved in a way that is scientifically sound, ethically responsible, and commercially viable. The integration of technology, collaborative decision-making, and patient-centric approaches further enhances the effectiveness of these collaborations, driving innovation and success in the pharmaceutical industry.

Through its role as a strategic collaborator, Medical Affairs contributes to the seamless coordination of medicinal product development and commercialization efforts, ensuring that products meet the needs of patients, HCPs, and regulators alike. As the pharmaceutical industry continues to evolve, the importance of cross-functional collaboration will only grow, solidifying Medical Affairs' role as a key driver of innovation and success.

Part II

Chapter 4
Modern Concepts and Strategy Development

Key Highlights
- Strategy aligns vision, mission, and long-term objectives with business operations.
- Situational analysis (internal and external) informs strategic decision-making.
- SWOT analysis identifies strengths, weaknesses, opportunities, and threats.
- Modern strategy concepts (Agile, Lean, Blue Ocean) emphasize flexibility and differentiation.
- Strategic execution includes resource allocation, monitoring, and iterative refinement.

In the context of business, strategy is a comprehensive plan formulated to achieve specific long-term goals and objectives. It serves as a roadmap that guides an organization in navigating complex and competitive environments to attain sustainable growth and success. Unlike short-term tactics or operational decisions that focus on immediate tasks, a strategy encompasses a broader vision that aligns with the organization's overall mission and values.

A well-developed strategy not only defines the direction in which an organization should move but also allocates the necessary resources—such as capital, personnel, and technology—to execute the plan effectively. It involves making deliberate choices about which opportunities to pursue and which to forego, based on an understanding of internal strengths and weaknesses as well as external opportunities and threats (often analyzed through frameworks like SWOT: Strengths, Weaknesses, Opportunities, Threats).

A. Krendyukov, *Medical Affairs' Fundamentals: The Next Generation*,
https://doi.org/10.1007/978-3-031-92588-7_4

Core Elements of Strategy Development

Vision and Mission

The starting point of any strategic process is the establishment of a clear vision and mission. The vision is a future-oriented declaration of the organization's purpose and aspirations—what it ultimately aims to achieve in the long term. The mission, on the other hand, is a concise explanation of the organization's reason for existence. It describes the core business activities and serves as a guide for decision-making, providing a path toward realizing the vision.

Strategic Objectives

Strategic objectives are specific, measurable goals that an organization sets to advance its mission and move toward its vision. These objectives must be clear, quantifiable, and time-bound. They provide a means for measuring progress and are often articulated in terms of key performance indicators (KPIs) that allow the organization to track its achievements against its goals.

Situational Analysis

A critical step in strategy development is conducting a situational analysis to understand the current state of the organization and its external environment. This involves the following:

Internal Analysis: Examining the organization's strengths and weaknesses, including resources, capabilities, and core competencies. This helps identify areas where the organization excels and where it may be vulnerable.

External Analysis: Assessing the opportunities and threats in the external environment, which includes market trends, competitive landscape, regulatory changes, and socio-economic factors. Tools like PESTEL analysis (Political, Economic, Social, Technological, Environmental, and Legal) are commonly used to evaluate these external factors.

SWOT Analysis (Fig. 4.1)

SWOT analysis—Strengths, Weaknesses, Opportunities, and Threats—stands out as one of the most versatile and widely used tools in this domain. By enabling organizations to comprehensively evaluate their internal and external environments, SWOT analysis serves as a key milestone in the formulation of effective strategies.

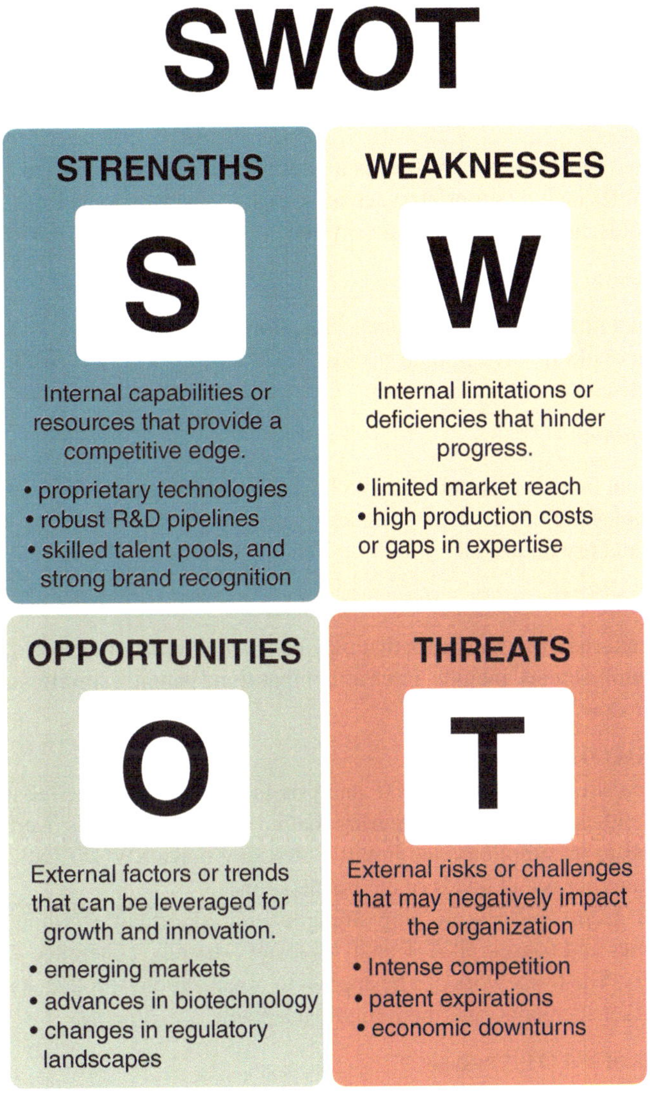

SWOT

STRENGTHS

S

Internal capabilities or resources that provide a competitive edge.

- proprietary technologies
- robust R&D pipelines
- skilled talent pools, and strong brand recognition

WEAKNESSES

W

Internal limitations or deficiencies that hinder progress.

- limited market reach
- high production costs or gaps in expertise

OPPORTUNITIES

O

External factors or trends that can be leveraged for growth and innovation.

- emerging markets
- advances in biotechnology
- changes in regulatory landscapes

THREATS

T

External risks or challenges that may negatively impact the organization

- Intense competition
- patent expirations
- economic downturns

Fig 4.1 SWOT

SWOT analysis is a structured planning method used to identify the internal strengths and weaknesses of an organization and the external opportunities and threats it faces. This four-quadrant framework provides decision-makers with a clear and holistic picture of the factors that influence their strategic objectives.

- *Strengths:*
    - Internal capabilities or resources that provide a competitive edge.
    - Examples include proprietary technologies, robust R&D pipelines, skilled talent pools, and strong brand recognition.
- *Weaknesses*:
    - Internal limitations or deficiencies that hinder progress.
    - These could involve limited market reach, high production costs, or gaps in expertise.
- *Opportunities*:
    - External factors or trends that can be leveraged for growth and innovation.
    - Examples include emerging markets, advances in biotechnology, or changes in regulatory landscapes.
- *Threats*:
    - External risks or challenges that may negatively impact the organization.
    - Common threats include intense competition, patent expirations, and economic downturns.

Values of SWOT Analysis

- *Comprehensive Overview*: SWOT analysis facilitates a balanced assessment of internal and external factors, ensuring that no critical element is overlooked.
- *Ease of Use*: Its simplicity and intuitive structure make SWOT analysis accessible to teams across all levels of an organization.
- *Strategic Alignment*: By aligning strengths and opportunities while mitigating weaknesses and threats, SWOT analysis fosters strategic coherence.
- *Versatility*: The framework can be applied to organizations, products, projects, or even market segments, making it highly adaptable.

Limitations of SWOT Analysis

- *Subjectivity*: Assessments within each quadrant can be subjective, influenced by individual biases or incomplete information.
- *Lack of Prioritization*: SWOT analysis does not inherently prioritize factors, leaving organizations to determine which elements are most critical.
- *Static Nature*: The analysis represents a snapshot in time and may quickly become outdated in dynamic environments.
- *Limited Strategic Guidance*: While SWOT identifies factors, it does not provide direct solutions or detailed action plans.

Applying SWOT Analysis in the Pharmaceutical Industry

The pharmaceutical sector is characterized by rapid innovation, stringent regulations, and fierce competition, making SWOT analysis a valuable tool for strategy formulation.

- *Strengths in Pharmaceuticals*: Robust intellectual property portfolios, innovative medicinal product pipelines, and strong collaborations with academic and research institutions are common strengths.
- *Weaknesses in Pharmaceuticals*: Long development cycles, high R&D costs, and dependency on successful clinical trials are typical challenges.
- *Opportunities in Pharmaceuticals*: Growth opportunities include expanding into emerging markets, leveraging artificial intelligence in innovative product discovery, and addressing unmet medical needs.
- *Threats in Pharmaceuticals*: Threats often stem from patent cliffs, evolving regulatory requirements, and increased competition from generics and biosimilars.

Using SWOT Analysis to Build Innovative Strategies

- *Match Strengths to Opportunities*: Leverage strong R&D capabilities to address opportunities in personalized medicine or rare disease treatments.
- *Address Weaknesses*: Invest in operational efficiency and diversify pipelines to mitigate vulnerabilities in product portfolios.
- *Monitor Threats*: Establish robust competitive intelligence systems to anticipate market disruptions and regulatory changes.
- *Develop Actionable Insights*: Combine SWOT analysis with other frameworks, such as PESTEL or Porter's Five Forces, to deepen the strategic context.

Tips for Effective SWOT Analysis

- *Engage Diverse Teams*: Involve cross functional teams to ensure a comprehensive and balanced analysis.
- *Quantify Where Possible*: Use metrics to substantiate claims, such as market share percentages or R&D investment figures.
- *Update Regularly*: Revisit SWOT analysis periodically to reflect evolving internal and external conditions.
- *Integrate with Strategic Planning*: Use SWOT findings as the foundation for setting objectives, allocating resources, and developing execution plans.

SWOT analysis remains a cornerstone in strategic planning due to its ability to provide a structured and comprehensive view of an organization's landscape. In the pharmaceutical industry, where innovation and adaptability are paramount, SWOT analysis can uncover critical insights that drive growth, mitigate risks, and enhance competitive positioning.

By understanding the strengths, limitations, and best practices of SWOT analysis, organizations can transform this simple framework into a powerful tool for navigating complexity and fostering long-term success. When combined with actionable strategies and an innovation-driven mindset, SWOT analysis serves not merely as a milestone but as a catalyst for transformative progress.

Strategic Formulation

Based on the insights gained from the situational analysis, organizations formulate strategies that leverage their strengths, mitigate their weaknesses, capitalize on opportunities, and defend against threats. This process may involve choosing from various strategic options such as market penetration, market development, product development, diversification, and strategic alliances.

Strategic Implementation

Once a strategy is formulated, the focus shifts to implementation, which involves translating strategic plans into actionable tasks. This phase requires allocating resources, assigning responsibilities, and setting timelines to ensure that strategic initiatives are executed effectively. It also involves establishing processes and systems to support implementation, such as organizational structures, communication channels, and performance management systems.

Monitoring and Evaluation

Continuous monitoring and evaluation are essential to assess the effectiveness of the strategy and ensure it is delivering the desired outcomes (refer Fig. 4.2). This involves tracking progress against strategic objectives and KPIs, analyzing performance data, and making adjustments as necessary. Regular reviews help organizations respond to changes in the internal and external environment and refine their strategies to maintain alignment with their goals.

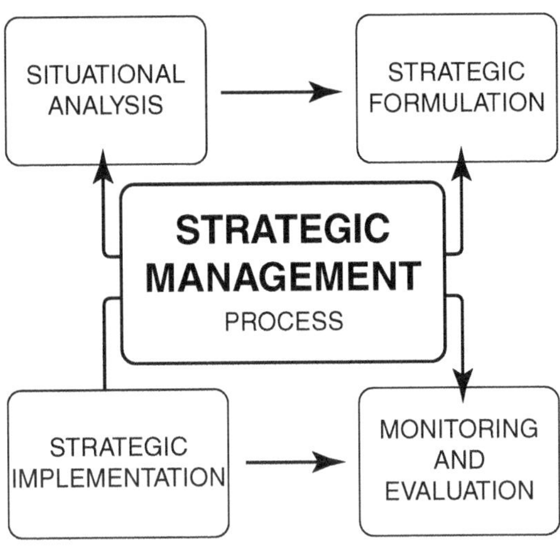

Fig. 4.2 Strategic management process

Essential Business Strategy Concepts

In an era marked by rapid advancements in science and technology, the pharmaceutical industry faces unprecedented opportunities and challenges. To thrive in this dynamic landscape, businesses must adopt robust strategies that enable sustainable growth, innovation, and resilience. Unfortunately, many organizations struggle to achieve a competitive edge due to unclear strategic direction or failure to leverage proven frameworks effectively. This section explores six essential business strategy concepts—the Ansoff Matrix, Porter's Five Forces, Value Disciplines, the Growth Share Matrix, P.E.S.T. Analysis, and Balanced Scorecard—through the lens of the pharmaceutical industry. Each framework is analyzed for its strengths, limitations, application in pharma, and tips for fostering innovative strategy development.

The Ansoff Matrix

The Ansoff Matrix is a strategic planning tool designed to identify opportunities for growth by examining market and product dimensions. It comprises four strategies: market penetration, product development, market development, and diversification. This model enables companies to evaluate risk levels and align their growth initiatives with corporate objectives.

Strengths

- *Simplicity*: The matrix provides a clear, visual framework for growth planning.
- *Versatility*: It accommodates various scenarios, from increasing market share to exploring new frontiers.
- *Risk Assessment*: By categorizing strategies by risk level, it supports informed decision-making.

Limitations

- *Oversimplification*: The matrix assumes static market conditions, which may not reflect real-world complexities.
- *Focus on Growth*: It may neglect operational and competitive dynamics.
- *Interdependencies*: Strategies often overlap, complicating execution.

Application for Pharma

Pharmaceutical companies can leverage the Ansoff Matrix to:

- *Penetrate Existing Markets*: Increase sales of existing drugs through targeted marketing and partnerships.
- *Develop New Products*: Invest in R&D to create innovative treatments for unmet needs.
- *Expand Into Untapped Markets*: Enter emerging markets where demand for healthcare solutions is growing.
- *Diversify Offerings*: Extend portfolios into adjacent areas, such as diagnostics, digital health, or wellness products.

Tips for Innovative Strategy Development

- Use real-world data to refine market penetration strategies, focusing on geographies with unmet needs.
- Combine product development with digital health tools, such as Artifical Intelligence (AI) for drug discovery or telemedicine platforms.
- Diversify with long-term sustainability in mind, balancing innovation with core business strengths.

Porter's Five Forces

Porter's Five Forces is a competitive analysis framework that evaluates industry attractiveness by examining five dynamics: the threat of new entrants, bargaining power of suppliers, bargaining power of buyers, the threat of substitutes, and competitive rivalry. It provides a holistic view of external pressures that shape strategy.

Strengths

- *Comprehensive View*: It examines multiple dimensions of competitive forces.
- *Industry Specificity*: Customizable to specific sectors, such as pharmaceuticals.
- *Strategic Insight*: Highlights potential risks and opportunities in the competitive landscape.

Limitations

- *Static Snapshot*: It may fail to account for rapidly changing conditions.
- *Internal Focus*: Does not address internal capabilities.
- *Complex Interactions*: Overlapping forces can complicate the analysis.

Application for Pharma

- *Threat of New Entrants*: Evaluate barriers such as regulatory approvals, capital requirements, and patent protections.
- *Supplier Power*: Assess relationships with Active Pharmaceutical Ingredient (API) suppliers, contract manufacturers, and raw material providers.
- *Buyer Power*: Address the influence of governments, insurers, and large health-care providers.
- *Substitution Risks*: Monitor the emergence of biosimilars and alternative therapies.
- *Competitive Rivalry*: Navigate intense competition in areas such as oncology, immunology, and rare diseases.

Tips for Innovative Strategy Development

- Strengthen barriers to entry by investing in intellectual property and regulatory expertise.
- Enhance supplier relationships through collaborative partnerships and supply chain diversification.
- Address buyer power by offering value-based pricing models and demonstrating clinical efficacy.

Value Disciplines

The Value Disciplines model identifies three strategic focuses for competitive differentiation: operational excellence, product leadership, and customer intimacy. Organizations are encouraged to excel in one area while maintaining baseline performance in the others.

Strengths

- *Clarity of Focus*: Encourages organizations to prioritize a core strength.
- *Flexibility*: Adaptable to diverse industries and business models.
- *Customer-Centric*: Aligns strategy with customer value.

Limitations
- *Trade-Offs*: Overemphasis on one discipline may lead to neglect of others.
- *Static Framework*: Less suited for rapidly evolving environments.
- *Execution Challenges*: Requires strong organizational alignment.

Application for Pharma
- *Operational Excellence*: Streamline manufacturing and distribution to reduce costs and improve efficiency.
- *Product Leadership*: Lead in innovation by developing groundbreaking therapies.
- *Customer Intimacy*: Build trust with patients and healthcare professionals through tailored services and patient support programs.

Tips for Innovative Strategy Development
- Leverage digital transformation to enhance operational excellence, such as using predictive analytics for supply chain optimization.
- Foster a culture of innovation to sustain product leadership in key therapeutic areas.
- Use patient-centric approaches to strengthen customer intimacy, including real-world evidence and personalized healthcare solutions.

Growth Share Matrix

The Growth Share Matrix, developed by the Boston Consulting Group (BCG), categorizes business units or products into four quadrants based on market growth and market share: Stars, Cash Cows, Question Marks, and Dogs. It guides resource allocation and portfolio management.

Strengths
- *Portfolio Management*: Simplifies resource allocation decisions.
- *Strategic Insight*: Highlights areas of investment, divestment, or optimization.
- *Focus on Return On Investment (ROI)*: Aligns resources with high-potential opportunities.

Limitations
- *Oversimplification*: Ignores market nuances and competitive dynamics.
- *Static Analysis*: Fails to capture market evolution over time.
- *Limited Scope*: Focuses only on market growth and share, neglecting other factors.

Application for Pharma
- *Stars*: High-growth, high-share areas such as cutting-edge therapies or blockbuster drugs.
- *Cash Cows*: Established drugs with steady demand, generating cash for R&D.
- *Question Marks*: Emerging markets or new products with uncertain potential.
- *Dogs*: Products or markets with low growth and limited share, potentially candidates for divestment.

Tips for Innovative Strategy Development

- Use predictive modeling to identify future stars, focusing on areas like cell and gene therapy.
- Reinvest Cash Cow profits into R&D and emerging markets to fuel long-term growth.
- Monitor external trends and pivot Question Marks into Stars or exit them strategically.

P.E.S.T. Analysis

P.E.S.T. Analysis examines macro-environmental factors that influence business: Political, Economic, Social, and Technological. It helps organizations understand external influences and adapt accordingly.

Strengths

- *Holistic View*: Captures external forces impacting strategy.
- *Scalability*: Applicable to various levels, from global to regional.
- *Proactive Planning*: Encourages anticipation of future trends.

Limitations

- *Broad Focus*: May overlook industry-specific factors.
- *Dynamic Changes*: Requires frequent updates to remain relevant.
- *Subjectivity*: Depends on accurate data interpretation.

Application for Pharma

- *Political*: Navigate regulatory landscapes and healthcare policies.
- *Economic*: Assess pricing pressures and funding challenges.
- *Social*: Address demographic shifts, patient expectations, and health equity.
- *Technological*: Leverage AI, big data, and advanced manufacturing technologies.

Tips for Innovative Strategy Development

- Monitor policy changes closely, ensuring compliance and adaptability.
- Analyze economic trends to develop cost-effective, scalable solutions.
- Invest in digital transformation to address both technological and social shifts.

Balanced Scorecard (BSC)

The Balanced Scorecard (BSC) is a performance management framework that aligns strategic objectives with measurable outcomes across four perspectives: financial, customer, internal processes, and learning and growth. It emphasizes a balanced approach to strategy execution, ensuring that financial goals do not overshadow other critical dimensions.

Strengths
- *Comprehensive Approach*: Integrates financial and non-financial metrics.
- *Alignment*: Links daily operations with long-term strategic goals.
- *Monitoring*: Provides a structured system for tracking progress.

Limitations
- *Implementation Complexity*: Requires a robust infrastructure for data collection and analysis.
- *Customization Needs*: Must be tailored to fit specific organizational goals.
- *Overemphasis on Metrics*: Risk of focusing too heavily on measurement rather than execution.

Application for Pharma
- *Financial Perspective*: Track revenue growth from new drugs and cost efficiency in production.
- *Customer Perspective*: Measure patient satisfaction, adherence rates, and market penetration.
- *Internal Processes*: Evaluate the efficiency of clinical trials, regulatory submissions, and manufacturing workflows.
- *Learning and Growth*: Focus on employee training, innovation in R&D, and fostering a collaborative culture.

Tips for Innovative Strategy Development
- Use the BSC to connect cross-functional goals, ensuring alignment between R&D, Marketing, Commercial and Regulatory teams.
- Incorporate real-world evidence (RWE) to enhance the customer perspective, emphasizing patient outcomes.
- Regularly review and adapt the BSC to reflect industry dynamics and organizational priorities.

The pharmaceutical industry's complex, rapidly evolving landscape demands sophisticated strategic approaches. The six frameworks discussed—the Ansoff Matrix, Porter's Five Forces, Value Disciplines, Growth Share Matrix, P.E.S.T. Analysis, and Balanced Scorecard—offer valuable tools for navigating challenges and seizing opportunities. By understanding their strengths, limitations, and applications, leaders can develop innovative strategies that align with their organization's goals. Ultimately, the key to success lies in combining these frameworks with forward-thinking, data-driven decision-making, and a commitment to delivering meaningful value to patients and stakeholders alike.

Modern Concepts in Strategy Development

The evolution of business environments, driven by globalization, technological advancements, and changing consumer behaviors, has led to the emergence of new concepts in strategy development. These modern approaches emphasize flexibility, agility, and a more iterative process compared to traditional, linear models (Table 4.1).

Table 4.1 Summary table on strategic frameworks, their benefits, and limitations in business and pharmaceutical company contexts

Strategy concept & framework	Brief description	Advantages & value	Limitations
The Ansoff Matrix	Identifies growth opportunities based on market and product dimensions (market penetration, product development, market development, and diversification).	—Simple and visual tool for growth planning.—Helps assess risk levels and align growth strategies.—Supports informed decision-making.	—Oversimplifies market conditions.—Focuses primarily on growth, neglecting operational and competitive factors.
Porter's Five Forces	Evaluates industry competitiveness through five forces: threat of new entrants, supplier power, buyer power, threat of substitutes, and competitive rivalry.	—Comprehensive view of competitive landscape.—Helps identify risks and opportunities.—Industry-specific customization possible.	—Provides a static snapshot, not reflecting rapid industry changes.—Does not assess internal capabilities.
Value Disciplines	Differentiation strategy based on excelling in one of three areas: operational excellence, product leadership, or customer intimacy.	—Provides clear strategic focus.—Aligns business strategy with customer needs.—Offers flexibility across industries.	—Overemphasis on one discipline can lead to neglect of others.—Requires strong organizational alignment.
Growth Share Matrix (BCG Matrix)	Categorizes business units/products into four quadrants based on market share and market growth: Stars, Cash Cows, Question Marks, and Dogs.	—Simplifies resource allocation and portfolio management.—Highlights investment, divestment, and optimization areas.	—Ignores market nuances and competitive dynamics.—Static analysis that may not capture market evolution.
P.E.S.T. Analysis	Assesses macro-environmental factors influencing business: political, economic, social, and technological.	—Provides a holistic external environment analysis.—Helps in proactive planning and market adaptation.	—Needs frequent updates to remain relevant.—May overlook industry-specific influences.
Balanced Scorecard (BSC)	Performance management framework aligning strategic objectives with measurable outcomes across financial, customer, internal process, and learning perspectives.	—Integrates financial and non-financial metrics.—Aligns operations with strategic goals.—Provides structured monitoring of performance.	—Complex implementation requiring significant data infrastructure.—Risk of excessive focus on metrics rather than execution.
Agile Strategy	An adaptive strategic approach emphasizing short planning cycles, iterative execution, and rapid response to change.	—Enhances flexibility in dynamic environments.—Encourages continuous improvement and learning.	—Requires cultural and structural adaptability.—May struggle with long-term vision and execution consistency.

(continued)

Table 4.1 (continued)

Strategy concept & framework	Brief description	Advantages & value	Limitations
Lean Strategy	Maximizing value while minimizing waste by streamlining processes and improving efficiency.	—Focuses on customer value and operational efficiency.—Encourages continuous process improvement.	—Can be difficult to sustain in complex industries.—May require cultural shifts for full implementation.
Blue Ocean Strategy	Focuses on creating new, uncontested market spaces instead of competing in existing saturated markets.	—Encourages innovation and differentiation.—Expands market reach and avoids price wars.	—Requires significant investment in innovation.—High risk if new markets do not develop as expected.
Disruptive Innovation	Introduces simpler, more affordable, or more accessible alternatives that challenge established market players.	—Can transform industries and create new value.—Provides opportunities for market entrants.	—Incumbent firms may struggle to adopt quickly.—Risk of unpredictable adoption and scalability.
VRIO Analysis	Evaluates firm resources based on value, rarity, imitability, and organization to assess sustainable competitive advantage.	—Helps identify and protect key strategic assets.—Encourages investment in long-term advantages.	—Subjective evaluation of resources.—Less effective in fast-changing industries.
The Value Stick	Framework optimizing value by managing the gap between Willingness to Pay (WTP) and Willingness to Sell (WTS).	—Ties strategy directly to customer and supplier perspectives.—Enhances margin optimization.	—Requires extensive market data for accurate assessment.—Assumes rational decision-making by stakeholders.
McKinsey's Strategic Horizons	Categorizes business activities into Horizon 1 (core business), Horizon 2 (emerging opportunities), and Horizon 3 (breakthrough innovations).	—Balances core operations with future innovation.—Ensures resource allocation for long-term growth.	—Challenging to manage across multiple horizons.—Risk of neglecting core business while pursuing innovation.
Strategy Diamond	Aligns strategic execution across five dimensions: arenas, vehicles, differentiators, staging, and economic logic.	—Provides comprehensive strategic alignment.—Ensures clarity across all strategic components.	—Complex implementation requiring frequent reviews.—May need adjustments in volatile markets.

Agile Strategy Development

Originally a concept from software development, agility has become a core principle in modern business strategy. Agile strategy development focuses on adaptability and responsiveness to change rather than strict adherence to a fixed plan. This approach is characterized by shorter planning cycles, iterative processes, and ongoing feedback loops that allow organizations to pivot quickly in response to new information or changing conditions.

Agile strategies are particularly valuable in industries where the pace of change is rapid and uncertainty is high, such as technology and digital services. By embracing agility, organizations can better manage risks, capitalize on emerging opportunities, and foster innovation. Agile frameworks like Scrum and Kanban, which emphasize collaboration, transparency, and continuous improvement, are commonly applied beyond Information Technology (IT) to strategic planning processes.

Lean Strategy

The lean approach, derived from lean manufacturing principles developed by Toyota, focuses on maximizing value while minimizing waste. Lean strategy involves continuously seeking ways to increase efficiency and effectiveness in all aspects of the organization, from product development to customer service. This is achieved by streamlining processes, eliminating non-value-adding activities, and optimizing the use of resources.

Lean strategy emphasizes a customer-centric approach, where value is defined from the customer's perspective. This focus on customer value ensures that strategic initiatives are aligned with customer needs and preferences, which is crucial in competitive markets. Lean tools such as value stream mapping, just-in-time (JIT) production, and continuous improvement (Kaizen) are integral to this approach.

Blue Ocean Strategy

The Blue Ocean Strategy, introduced by W. Chan Kim and Renée Mauborgne, is based on the idea of creating new, uncontested market spaces—"blue oceans"—rather than competing in overcrowded industries or "red oceans." In a red ocean, companies fight over a shrinking profit pool through intense competition, often leading to commoditization and price wars. In contrast, blue ocean strategy focuses on innovation and differentiation to create new demand and unlock new growth opportunities.

This strategy involves redefining market boundaries, creating unique value propositions, and pursuing differentiation and low cost simultaneously. By doing so, organizations can tap into new customer segments and make competition irrelevant. A famous example is the Nintendo Wii, which attracted non-gamers by offering a unique, physically interactive gaming experience, effectively expanding the market.

Disruptive Innovation

Coined by Clayton Christensen, disruptive innovation describes a process where smaller companies, often with fewer resources, successfully challenge established incumbents by introducing simpler, more affordable, or more convenient alternatives. These innovations often begin by targeting underserved or overlooked segments of the market that established players have ignored. Over time, as the innovation improves and gains wider adoption, it can disrupt and eventually displace the incumbent technologies or business models.

A classic example is the rise of digital streaming services, which disrupted traditional cable television. Companies like Netflix and Hulu started by offering lower-cost alternatives with on-demand viewing, appealing to consumers dissatisfied with conventional models. As the platforms grew and refined their offerings, they attracted mainstream consumers and fundamentally changed the way media content is consumed.

Disruptive innovation is especially important in industries with high barriers to entry, such as pharmaceuticals, where established companies dominate. In other sectors, like technology, disruptors are reshaping markets and setting new industry standards by leveraging digital platforms, automation, and AI.

Additional Strategic Frameworks

Developing effective strategies in the pharmaceutical business requires the integration of robust analytical frameworks to uncover competitive advantages, optimize value creation, and balance core operations with innovative breakthroughs. This section explores four strategic frameworks—VRIO Analysis, the Value Stick, McKinsey's Strategic Horizons, and the Strategy Diamond—highlighting their strengths, limitations, and applicability in fostering innovation and sustaining competitive advantage in the pharmaceutical sector.

VRIO Analysis *Assessing Resources for Sustained Competitive Advantage*

The VRIO framework, developed by Barney, is a tool for evaluating a firm's resources and capabilities. It identifies whether these assets provide a sustained competitive advantage by analyzing four dimensions: value, rarity, imitability, and organization.

- *Overview:*
  - Value: Does the resource enable the firm to exploit opportunities or neutralize threats?
  - Rarity: Is the resource scarce among competitors?
  - Imitability: Is it costly for others to imitate or substitute?
  - Organization: Is the firm organized to capture value from the resource?
- *Strengths*:
  - Offers a structured approach to identify key resources.
  - Focuses on long-term competitive sustainability.
- *Limitations*:
  - Subjective assessment of resources.
  - Less effective in dynamic environments with rapid technological changes.
- *Application in Pharmaceuticals*:
  - Identify proprietary drug formulas, patents, or unique R&D capabilities as valuable and rare resources.
  - Invest in protecting these resources through robust legal and organizational strategies to enhance their inimitability.
- *Tips for Innovative Strategy*:
  - Conduct periodic VRIO analyses to adapt to evolving market conditions.
  - Focus on building capabilities in advanced technologies like AI-driven drug discovery.

The Value Stick *Balancing Willingness to Pay and Willingness to Sell*

The Value Stick, conceptualized by Felix Oberholzer-Gee, emphasizes maximizing customer delight and firm margins by managing the two ends of the value spectrum—Willingness to Pay (WTP) and Willingness to Sell (WTS).

- *Overview*:
  - *WTP*: The maximum price a customer is willing to pay.
  - *WTS*: The minimum price at which a supplier is willing to sell.
  - The gap between WTP and WTS represents the value created.
- *Strengths*:
  - Directly ties strategy to customer and supplier perspectives.
  - Provides actionable insights to enhance margins and customer satisfaction.
- *Limitations*:
  - Requires detailed market data to accurately assess WTP and WTS.
  - Assumes rational decision-making by stakeholders.

- *Application in Pharmaceuticals*:
  - Increase WTP by investing in patient-centric innovations, such as personalized medicine or enhanced drug delivery mechanisms.
  - Reduce WTS by streamlining supply chains and leveraging economies of scale in manufacturing.
- *Tips for Innovative Strategy*:
  - Collaborate with patient advocacy groups to understand unmet needs and enhance WTP.
  - Negotiate strategic partnerships with suppliers to optimize WTS.

McKinsey's Strategic Horizons *Balancing Core Business and Innovation*

The Strategic Horizons model categorizes business activities into three horizons—Horizon 1 (Core Business), Horizon 2 (Emerging Opportunities), and Horizon 3 (Breakthrough Innovations). It ensures a balanced approach to sustaining core operations while exploring future opportunities.

- *Overview*:
  - *Horizon 1:* Focuses on sustaining and growing the core business.
  - *Horizon 2*: Explores adjacent opportunities and new market segments.
  - *Horizon 3*: Pursues transformative innovations and disruptive technologies.
- *Strengths*:
  - Encourages diversification without neglecting core operations.
  - Provides a framework for resource allocation across timelines.
- *Limitations*:
  - Balancing resources across horizons can be challenging.
  - Requires robust governance structures to avoid neglecting any horizon.
- *Application in Pharmaceuticals*:
  - *Horizon 1*: Focus on lifecycle management of existing blockbuster drugs.
  - *Horizon 2*: Invest in biosimilars or expand into emerging markets.
  - *Horizon 3*: Develop transformative therapies like gene editing or mRNA-based treatments.
- *Tips for Innovative Strategy*:
  - Allocate a specific percentage of R&D budgets to each horizon.
  - Foster a culture of innovation by incentivizing breakthrough research.

Strategy Diamond *Aligning Elements for Strategic Execution*

Developed by Hambrick and Fredrickson, the Strategy Diamond offers a holistic view of strategy by aligning five interrelated elements—Arenas, Vehicles, Differentiators, Staging, and Economic Logic.

- *Overview*:
  - *Arenas*: Where will the company compete (e.g., markets, products)?
  - *Vehicles*: How will the company get there (e.g., partnerships, acquisitions)?
  - *Differentiators*: How will the company win (e.g., unique value proposition)?
  - *Staging*: What will be the sequence and pace of moves?
  - *Economic Logic*: How will the company generate profits?
- *Strengths*:
  - Provides a comprehensive framework for strategic alignment.
  - Ensures clarity and coherence across all strategic dimensions.
- *Limitations*:
  - Complexity in aligning all elements.
  - May require frequent adjustments in volatile markets.
- *Application in Pharmaceuticals*:
  - Define arenas by targeting specific therapeutic areas or patient demographics.
  - Choose vehicles such as strategic alliances with biotech firms for innovation.
  - Differentiate through unique drug efficacy profiles or patient support programs.
- *Tips for Innovative Strategy*:
  - Regularly revisit the strategy diamond to ensure alignment with market dynamics.
  - Use RWE and patient outcomes to refine differentiators.

Strategic frameworks like VRIO Analysis, the Value Stick, McKinsey's Strategic Horizons, and the Strategy Diamond offer valuable tools for building robust, innovative product strategies in the pharmaceutical sector. Each framework provides distinct insights—from resource evaluation to value optimization and balancing innovation with core business priorities.

By leveraging these additional frameworks effectively, pharmaceutical companies can enhance their competitive positioning, drive patient-centric innovation, and achieve sustainable growth. Combining the strengths of multiple frameworks and tailoring them to specific organizational needs ensures a dynamic and resilient strategy, paving the way for transformative advancements in healthcare.

Strategy Development in the Pharmaceutical Environment

While modern business concepts such as Agile, Lean, Blue Ocean strategy, and Disruptive Innovation provide useful frameworks for strategy development, their direct application in the pharmaceutical industry requires significant adaptation.

The pharmaceutical sector, and oncology in particular, operates in a highly regulated, scientifically driven environment with longer product development cycles and unique pricing challenges.

Pharmaceutical strategy development is less consumer-centric and more product-centric, with a strong emphasis on scientific innovation, regulatory compliance, and value-based pricing. In contrast to the rapid iterations seen in technology or consumer goods, pharmaceutical companies must navigate extended R&D timelines and prioritize safety, efficacy, and market access in their strategic planning.

As pharmaceutical companies continue to innovate in areas like oncology, they must also adapt modern business strategies to fit their unique environment, integrating scientific rigor with the agility and flexibility required to meet the evolving demands of healthcare systems and patients.

Case Study: Development of Innovative Medicinal Products (InMPs) in Oncology *(based on the published paper A. Krendyukov & D. Nasy "Medical Affairs and Innovative Medicinal Product Strategy Development", Pharmaceutical Medicine, 2022)*

The global population is expanding, aging, and becoming increasingly sedentary. These trends contribute to a rise in the incidence of chronic diseases, such as cancer, which have high mortality rates. At the same time, improvements in diagnostic methods are allowing for earlier disease detection, often at more curable stages. As more people gain access to affordable long-term healthcare, particularly in developed countries, the demand for innovative medicinal products (InMPs) that improve the quality of life and extend survival has grown.

However, the development and commercialization of InMPs are marked by complex challenges, especially in the oncology field. Unlike other industries, pharmaceutical companies must navigate stringent regulatory environments, long development timelines, and unpredictable clinical outcomes. Strategic frameworks borrowed from non-pharmaceutical sectors like Fast-Moving Consumer Goods (FMCG) or the automobile industry are often inappropriate or require significant adaptation to suit the unique demands of pharmaceutical innovation. This is particularly true in oncology, where the heterogeneity of cancer mutations necessitates highly specialized treatment approaches.

In this context, it is critical for pharmaceutical companies to identify specific opportunities for innovation within particular therapeutic areas and align their product development strategies to maximize impact. An integrated strategy must encompass the entire life cycle of an InMP—from early research through clinical trials to post-market surveillance and commercialization. Successful InMPs must also address the needs of all stakeholders, including patients, healthcare professionals, regulatory bodies, and payers, while adapting to the evolving healthcare landscape.

Factors Influencing InMP Strategy Development

High Research and Development (R&D) Costs

Unlike non-pharmaceutical sectors, where product development often begins with an understanding of consumer needs, pharmaceutical innovation generally starts in the laboratory with the discovery or synthesis of an active compound. The journey from discovery to market is long and fraught with uncertainties (Fig. 4.3). On average, it takes 12–13 years for an InMP to reach the market, with escalating R&D costs along the way. In oncology, these challenges are compounded by high failure rates in clinical trials and stringent regulatory requirements.

Even with regulatory incentives such as the FDA's breakthrough therapy designation or the EMA's priority medicine (PRIME) designation, companies may struggle to recover R&D costs before patent protections expire or competitors bring similar products to market faster. Thus, the financial risks of InMP development are significant, and pricing strategies must account for the high cost of failure and lengthy development timelines.

Cost and Profitability

The financial burden of cancer treatment continues to grow. Global spending on cancer medicine increased to $223 billion in 2023 and is projected to reach $409 billion by 2028. The mean cost to deliver a new oncology medicine was recently estimated as $4.4b (95% CI, $3.6–5.2b), with a likelihood of approval of 5.3% for oncology InMPs entering clinical development.

This high upfront investment and the long lag before returns on investment require companies to set high launch prices for new InMPs. For example, the FDA's approval of CAR T-cell therapies like tisagenlecleucel and axicabtagene ciloleucel set new benchmarks for cancer drug pricing. While high prices are necessary to sustain innovation, they also raise concerns about affordability and access to essential treatments, leading to calls for strategies that balance innovation with affordability.

Complex Legislation

In contrast to other industries, the pharmaceutical sector is heavily regulated, with every stage of InMP development subject to detailed legislative requirements. Oncology trials, in particular, are longer and more complex than those in other therapeutic areas. The average oncology trial lasts 3.2 years, compared to 1.8 years for other therapy areas.

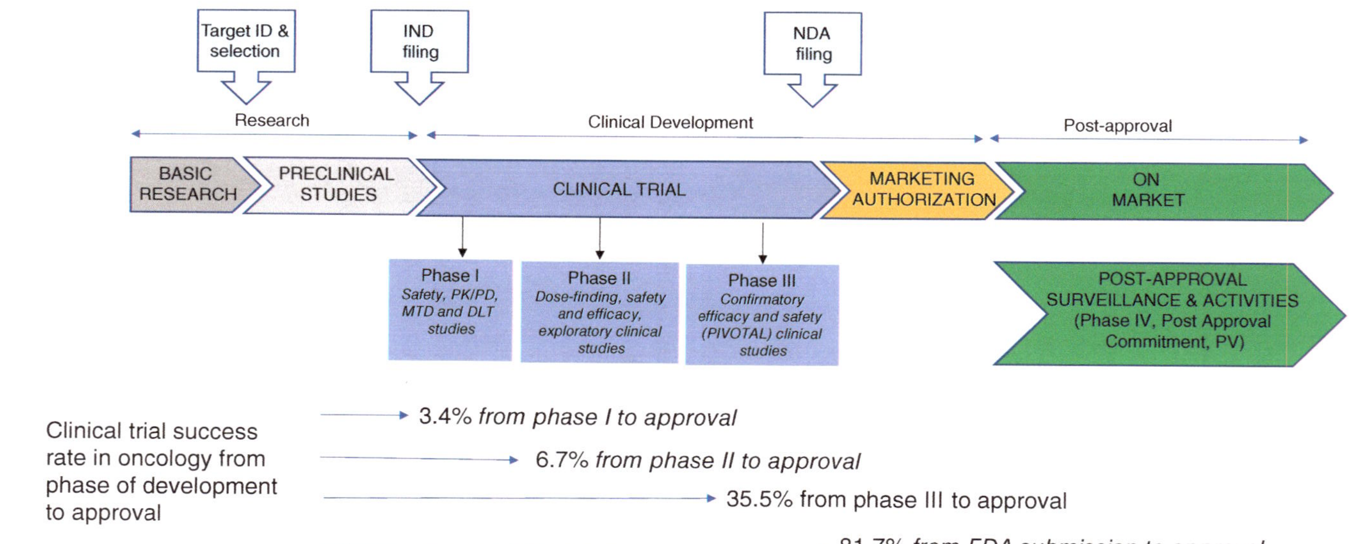

FDA, Food and Drug Administration; IND, Investigational New Drug; NDA, New Drug Application; PV, Pharmacovigilance; DLT, Dose L imiting Toxicity; MTD, Maximum Tolerated Dose .

Fig. 4.3 Drug discovery process from target identification and validation through to filing of a compound and the approximate success rates for these processes

Expedited regulatory pathways, such as orphan drug designations, can shorten review times but do not necessarily reduce clinical trial durations. As regulatory bodies like the FDA continue to streamline approval processes for oncology drugs, companies must develop strategies that navigate complex legislation while maintaining compliance with evolving regulations.

Pricing and Reimbursement

The pharmaceutical industry is increasingly scrutinized for the high cost of treatments, particularly in oncology. Unlike other sectors, where consumers directly pay for products, InMPs are typically covered by a combination of insurance and government reimbursement schemes. Health Technology Assessments (HTAs) are becoming standard practice in Europe and the United States to determine the cost-effectiveness of new therapies.

Payers are adopting new strategies to negotiate prices based on the value of InMPs. This includes incorporating outcomes like overall survival and quality of life into reimbursement decisions, placing greater pressure on pharmaceutical companies to demonstrate the real-world value of their products.

Broad Customer Base

The complexity of decision-making in healthcare further complicates InMP strategy. Unlike other industries where the end consumer is often the primary decision-maker, multiple stakeholders—including regulatory bodies, healthcare providers, insurers, and patients—are involved in InMP evaluation and approval. Each group has distinct needs and influences, making it challenging to balance product attributes with stakeholder expectations.

InMP Strategy: Product-Centric or End-Consumer-Centric?

In industries outside of healthcare, product development is driven by consumer needs. Successful product strategies focus on differentiation and competitive advantage, addressing customer demands at every stage of the product life cycle. This contrasts with the pharmaceutical industry, where product attributes often drive development, rather than consumer needs.

In oncology, the focus is on the product's scientific and clinical attributes, rather than direct consumer needs. The identification of biomarkers and the selection of patient subpopulations for clinical trials are critical to demonstrating the efficacy of

InMPs. For instance, companion diagnostic tests are crucial for identifying patients who will benefit most from specific therapies.

While patient needs are important, it is product-specific attributes that drive success in this industry. Pharmaceutical companies must focus on understanding the disease pathway, clinical outcomes, and cost-effectiveness to maximize the value of their InMPs.

Evaluation and Approval of InMPs

The evaluation and approval of InMPs are complex processes involving multiple stakeholders. In oncology, value-based healthcare is becoming a key consideration, with frameworks like those from the American Society of Clinical Oncology (ASCO) and the European Society for Medical Oncology (ESMO) guiding pricing and reimbursement decisions based on clinical outcomes, toxicity, and quality of life.

Strategic Focus in Post-Authorization

After InMP approval, pharmaceutical companies must engage stakeholders through evidence-based communication. Social media and Medical Affairs teams play a key role in sharing clinical trial results and engaging with physicians and patients. In oncology, post-authorization strategies must focus on maintaining stakeholder engagement and leveraging real-world data to improve product positioning.

Current Trends Shaping InMP Strategy in Oncology

The rise of precision medicine, new clinical trial designs, and advancements in AI and digital therapeutics are transforming oncology InMP development. Basket and umbrella trials, decentralized clinical trials, and partnerships with genetic testing companies are helping to streamline product development and improve patient outcomes.

AI is playing a crucial role in oncology, from guiding clinical decision-making to predicting treatment responses. As these technologies continue to evolve, they will reshape the way InMP strategies are developed and implemented.

Future Considerations in InMP Strategy Development

In the future, InMP strategies must adapt to evolving payment models, such as bundled payments in the US healthcare system, where providers are compensated based on patient outcomes. This shift towards value-based healthcare will place greater emphasis on the clinical benefits of InMPs and the ability of companies to demonstrate their value through integrated product strategies.

Pharmaceutical companies must also embrace new business models, such as outcome-based pricing, to ensure their products remain accessible and affordable while driving innovation in cancer care.

The development of InMPs in oncology requires a unique approach that differs from traditional business models. The success of an InMP is dependent on:

- A durable, integrated product strategy
- A solid clinical evidence base
- Effective differentiation from available therapeutic alternatives

To succeed, pharmaceutical companies must focus on developing strategies that address the needs of all stakeholders, incorporating clinical benefit scales and patient-reported outcomes from the outset to maximize the chances of success.

Chapter 5
Scientific Communication in Pharma

Key Highlights

- Scientific communication ensures accurate dissemination of clinical and medical information.
- Publication strategies align with regulatory submissions and key data milestones.
- Digital transformation enhances engagement via webinars, virtual congresses, and open access.
- Ethical guidelines (ICMJE, GPP4) govern transparency and responsible reporting.
- Addressing misinformation and aligning global communication with regulatory standards is crucial.

In today's pharmaceutical industry, the importance of scientific communication has never been greater. The rapid pace of scientific advances, coupled with a more globally interconnected healthcare environment and increasing scrutiny from regulators, necessitates that pharmaceutical companies develop comprehensive, transparent, and strategic communication plans. The communication of scientific findings, clinical trial data, and therapeutic advancements must not only meet the rigorous standards of accuracy and ethics but also adapt to modern trends such as digital dissemination, patient-centric engagement, and global regulatory harmonization.

Scientific communication plays a crucial role in ensuring that the research, development, and clinical outcomes generated by pharmaceutical companies are disseminated to the relevant audiences. It differs from corporate and promotional communication in both purpose and approach. In this section, we will define scientific communication, its scope, and its distinction from other types of communication within pharma companies.

A. Krendyukov, *Medical Affairs' Fundamentals: The Next Generation*,
https://doi.org/10.1007/978-3-031-92588-7_5

Some Definitions

Scientific (Non-Promotional) Communication: Involves the dissemination of peer-reviewed research, clinical trial data, and other scientific findings to healthcare professionals, regulatory bodies, and the broader scientific community. It is focused on accuracy, transparency, and credibility.

Corporate Communication: Primarily focuses on brand positioning, business updates, investor relations, and broader company news. The tone is often more marketing-oriented, and the audience is diverse, including investors, employees, and the general public.

Promotional Communication: Intended to drive product sales and brand recognition. This includes advertising, sales collateral, and direct marketing to healthcare providers (HCPs) and patients. Unlike scientific communication, it often follows stricter legal and regulatory constraints, such as being compliant with country-specific promotional codes.

Refining the Introduction for Greater Context

Scientific communication is the cornerstone of pharmaceutical research, development, and clinical practice, ensuring that HCPs, regulators, and other stakeholders are equipped with the information necessary to make informed decisions about therapies and treatments. The evolution of this field mirrors the advances in medicine and biotechnology, making it increasingly vital for pharmaceutical companies to develop strategic communication plans.

The Growing Complexity of Scientific Communication in Pharma

In the era of rapid scientific advancement, especially with the rise of biologics, biosimilars, and personalized medicine, the role of scientific communication has expanded significantly. Biologics, complex molecules derived from living organisms, and biosimilars, which are highly similar but not identical to these originator biologics, have introduced new challenges in explaining their development, efficacy, and safety. These therapies often require more detailed and precise communication with HCPs due to their inherent complexity, regulatory scrutiny, and public concern.

Personalized medicine, which tailors treatment to individual patient characteristics such as genetics, further intensifies the need for precise communication strategies. As healthcare systems adopt these treatments, it is crucial for pharmaceutical companies to provide transparent, evidence-based information to guide HCPs in selecting appropriate therapies. Scientific communication serves not only to inform but also to clarify complex therapeutic concepts and dispel misconceptions, such as those related to biosimilar safety or efficacy compared to reference products.

Building Trust and Preventing Misinformation

Pharmaceutical companies are under increasing pressure to maintain the integrity of their communications. The potential for misinformation—whether through misunderstanding of scientific concepts or intentional misrepresentation—can undermine trust in both specific treatments and the broader pharmaceutical industry. In this environment, clear, transparent scientific communication is essential for building trust with HCPs and the public. Well-structured communication strategies ensure that accurate information reaches the right audience, helping to mitigate the spread of misinformation and ensuring informed decision-making in clinical practice.

The Concept of Publication Strategy and Plans

A robust scientific communication strategy often begins with a clear publication plan, outlining how and when to disseminate clinical data and other research findings.

ICMJE Authorship Criteria and GPP4 Guidelines

The International Committee of Medical Journal Editors (ICMJE) guidelines are widely accepted for determining authorship in biomedical publications. These guidelines ensure that contributors who meet specific criteria are listed as authors, providing transparency and accountability.

The GPP4 guidelines (Good Publication Practice for Communicating Company-Sponsored Medical Research) further ensure ethical and transparent practices in the dissemination of research by pharmaceutical companies.

Overall Timelines for Manuscript Development

Publication timelines should be synchronized with key clinical and regulatory milestones. Typically, data is published soon after regulatory submission or approval.

Manuscript development should follow a structured process, with predefined timelines for drafting, internal review, and submission to target journals.

Clinical Data Reporting and Transparency

Reporting clinical data requires adherence to transparency regulations such as ClinicalTrials.gov and the EU Clinical Trials Register, ensuring that data is made publicly available in a timely manner.

Synchronizing the publication of clinical trial results with regulatory milestones ensures both transparency and trust in the company's scientific rigor.

Pitfalls to Avoid

Challenges that often arise in scientific communication strategies can delay or disrupt the process:

Too Many Co-authors: Having an excessive number of co-authors can slow down the review process. Clear definitions of each author's role are essential to maintain efficiency.

External or Dedicated Project Manager: It is recommended to have a dedicated person, preferably external to the company, to manage the process. This helps in unbiased communication and better organization.

Role of Lead/Senior Authors: Defining the role of lead authors helps to resolve disputes or ambiguities quickly. They should take ownership of the manuscript and act as decision-makers in case of disagreements.

Comment Alignment and Dispute Resolution: It is crucial to resolve comments or feedback efficiently. For instance, categorizing feedback as minor or major helps in prioritizing and resolving disagreements among authors.

Expanding on Publication Strategy to Include Newer Trends

The landscape of scientific publication is rapidly changing, and pharmaceutical companies must keep pace with these transformations to effectively disseminate their findings. Traditional publication models, while still relevant, are being complemented—and in some cases, replaced—by newer trends that offer faster, more accessible ways to share scientific data.

Open Access Publications

The rise of open access publishing has been one of the most significant developments in scientific communication in recent years. Unlike traditional subscription-based models, open access journals allow unrestricted access to published research, making scientific findings more widely available to both the scientific community and the public. For pharmaceutical companies, open access publications offer an opportunity to demonstrate their commitment to transparency and accessibility by ensuring that important research findings are not hidden behind paywalls.

Open access also aligns with the growing demand for patient access to scientific information, as more patients become involved in their own healthcare decisions. By choosing open access routes, pharmaceutical companies can ensure that key stakeholders—including patients, advocacy groups, and under-resourced medical communities—have equitable access to cutting-edge scientific data.

Preprint Platforms for Rapid Data Dissemination

Another important trend is the growing use of preprint platforms. Preprints are manuscripts that are made publicly available before undergoing formal peer review, allowing for rapid dissemination of research findings. This has been particularly impactful during times of crisis, such as the COVID-19 pandemic, when the need for immediate access to clinical data was critical. Preprints provide a mechanism for pharmaceutical companies to share important findings quickly, enabling faster feedback from the scientific community and, potentially, earlier clinical application of new therapies.

However, preprints also come with challenges. The lack of peer review means that companies must be cautious to ensure that data is presented responsibly and that limitations are clearly communicated to avoid misinterpretation. Nevertheless, the use of preprints is becoming a key element of modern scientific communication strategies, particularly for data that is time-sensitive or critical to public health.

What is Medical Communication?

Medical communication focuses on engaging HCPs, including physicians, nurses, pharmacists, and medical societies. It aims to ensure that HCPs are informed about new drugs, clinical guidelines, and medical advancements.

Key Focus: Medical communication highlights the clinical benefits, safety, and usage of pharmaceutical products.

Target Audience: It is directed toward HCPs who make prescribing decisions, as well as regulatory bodies.

Venues: Medical communication takes place through channels such as Continuing Medical Education (CME) programs, journal publications, and educational symposia.

Integrating Digital Channels in Medical Communication

The digital transformation of medical communication has accelerated in recent years, spurred in part by the COVID-19 pandemic. With travel restrictions and the shift to remote working practices, traditional venues for medical communication, such as in-person congresses and symposia, have increasingly moved online.

Virtual Congresses and Symposia

Virtual congresses and digital symposia have become essential tools for pharmaceutical companies to engage with HCPs in the absence of in-person events. These digital platforms offer several advantages, including the ability to reach a global audience, reduced costs, and the potential for extended engagement beyond the congress itself, as content can be accessed on demand.

Pharmaceutical companies must optimize their strategies for engaging HCPs through digital platforms. This includes creating interactive, high-quality content that can capture the attention of virtual attendees, such as live-streamed presentations, on-demand video sessions, and real-time Q&A sessions with experts. Effective digital communication also requires the use of metrics to assess engagement and adjust strategies accordingly, ensuring that the scientific message is reaching its intended audience.

Webinars and Online Continuing Medical Education (CME)

The use of webinars and online CME programs has also seen significant growth. These platforms allow companies to offer educational content to HCPs, providing updates on new therapies, clinical guidelines, and emerging research. In addition to serving as a valuable tool for disseminating scientific information, webinars and online CME programs allow for continuous engagement with HCPs, fostering ongoing dialogue and learning.

In this digital era, pharmaceutical companies must integrate these tools into their broader communication strategies, ensuring that they not only convey scientific information effectively but also maintain regulatory compliance in a digital environment.

Congress Presence: Booths and Other Activities

Scientific congresses and conferences are pivotal for pharmaceutical companies to showcase new research findings, interact with thought leaders, and generate awareness about new developments.

Booth Activities: The company's congress booth provides an opportunity for direct interaction with attendees, including clinicians, researchers, and policymakers. It serves as a hub for scientific exchange and engagement.

Posters (and Oral Presentation): Poster sessions offer a platform for presenting data that may not yet be ready for a full publication. They provide an informal setting to discuss findings and gather feedback from peers.

Advisory Boards (Ad Boards): Advisory boards are often held in conjunction with major congresses, allowing the company to gather insights from key opinion leaders (KOLs) on product development or clinical strategies.

Symposia or Colloquia: Key Differences

While both symposia and colloquia are platforms for scientific discussion, they have key distinctions:

Symposia: Typically more formal and structured, symposia involve invited speakers presenting on a specific theme, followed by a question-and-answer session. It is an excellent venue for sharing detailed clinical trial results or discussing advancements in therapeutic areas.

Colloquia: These are often smaller, more interactive meetings where participants engage in open discussions. Colloquia provide a space for in-depth dialogue on emerging issues, trends, or specific challenges in a therapeutic area.

More Emphasis on Globalization of Communication

As pharmaceutical companies expand their operations globally, the challenges of scientific communication become more complex. The need for consistent, transparent communication across multiple regions, each with its own regulatory environment, adds layers of complexity to the development and execution of communication strategies.

Navigating Diverse Regulatory Requirements

Regulatory frameworks such as those of the FDA in the United States, the EMA in Europe, and the PMDA in Japan have unique requirements for scientific communication, particularly regarding the reporting of clinical trial data, adverse events, and promotional activities. Pharmaceutical companies must navigate these differing regulations while maintaining a unified communication strategy that resonates across regions.

This requires close coordination between global and local teams, as well as a deep understanding of regional healthcare environments. Tailoring communication to the specific needs and expectations of local regulators, HCPs, and patients is essential for ensuring the successful dissemination of scientific information. This is particularly important when communicating complex scientific concepts, such as the development and approval of biosimilars, where understanding regulatory nuances is key to building trust and adoption across markets.

Adapting to Cultural Differences

Beyond regulatory considerations, companies must also consider cultural differences in healthcare communication. The way scientific data is presented and received can vary widely between regions. For instance, some regions may prioritize certain safety data, while others may focus more on efficacy or cost-effectiveness. A successful global communication strategy must be flexible enough to accommodate these cultural differences while maintaining consistency in the core scientific message.

Strengthening the Ethical Perspective

The pharmaceutical industry operates at the intersection of science, healthcare, and business, making it essential to maintain the highest ethical standards in scientific communication. The tension between commercial interests and the need for objective, transparent reporting of scientific data can create ethical challenges, particularly in the dissemination of clinical trial results.

One of the most significant ethical challenges in scientific communication is the risk of selective reporting. This occurs when companies choose to publish only positive results, leaving out negative or inconclusive findings that may provide important context for understanding a therapy's overall safety and efficacy. Such practices can erode trust with HCPs, regulators, and the public.

To combat selective reporting, pharmaceutical companies must commit to publishing all clinical trial data, regardless of the outcome, and adhere to guidelines such as the Good Publication Practice (GPP4) standards. This level of transparency is critical for ensuring that HCPs can make fully informed decisions based on a comprehensive view of the available evidence.

Patient-Centric Communication as an Emerging Trend

The pharmaceutical industry is increasingly adopting a patient-centric approach, recognizing that patients themselves are becoming more engaged in their healthcare decisions. This shift is also reflected in scientific communication strategies, with companies striving to make their research more accessible and understandable to non-expert audiences.

Pharmaceutical companies are beginning to work directly with patient advocacy groups to create patient-friendly summaries of clinical trial results, making complex scientific data more accessible to the general public. These summaries often highlight the key findings in simple, non-technical language, addressing concerns that matter most to patients, such as treatment options, side effects, and potential outcomes.

Moreover, there is a growing trend toward involving patients in the clinical trial process itself, from the design stage to post-trial communication. By giving patients a voice in how scientific data is communicated, pharmaceutical companies can foster greater trust and collaboration, ensuring that their communications resonate not only with HCPs but also with the patients who will ultimately benefit from these therapies.

Conclusion

In the pharmaceutical industry, a well-designed scientific communication strategy is essential for ensuring the successful dissemination of research findings and clinical trial data to relevant audiences. It requires careful planning, adherence to ethical guidelines, and synchronization with clinical and regulatory milestones. Effective scientific communication not only helps educate HCPs but also contributes to the transparency and credibility of the pharmaceutical company.

Scientific communication in the pharmaceutical industry is evolving rapidly, driven by advances in medicine, digital transformation, and the increasing globalization of healthcare. By refining their communication strategies to incorporate newer trends such as open access publishing, digital platforms, and patient-centric approaches, pharmaceutical companies can ensure that they are effectively reaching their target audiences. At the same time, maintaining a strong ethical foundation and navigating the complexities of global communication are critical to preserving trust and transparency. As the industry continues to evolve, so too must its communication strategies, ensuring that scientific data is disseminated responsibly, transparently, and ethically to support HCPs, patients, and regulators alike.

Tips for Analyzing and Presenting Results from Systematic Reviews and Meta-Analyses

Systematic reviews (SRs) and meta-analyses (Table 5.1) are cornerstone methodologies in evidence-based medicine. They synthesize existing research, enabling clinicians and researchers to make informed decisions grounded in a robust aggregation of data. This section elucidates the key elements involved in analyzing and presenting results derived from these studies, ensuring clarity, reproducibility, and utility in clinical and scientific applications.

Table 5.1 Main types of literature reviews

Types	Definition	Focus
Systematic review	A structured approach to collecting, analyzing, and synthesizing studies.	Answering a specific question with comprehensive and unbiased coverage.
Integrative review	Combines and critiques existing research to provide a broad understanding.	Identifying gaps and synthesizing knowledge across studies.
Theoretical review	Examines models, frameworks, and concepts to refine or propose theories.	Developing hypotheses and advancing theoretical understanding.
Argumentative review	Evaluates selected literature to support or refute a specific perspective.	Critical analysis to justify or challenge viewpoints.
Historical review	Explores the evolution of a topic or concept over time.	Contextualizing current knowledge within a historical timeline.
Scoping review	Maps key concepts, evidence types, and gaps in research on a topic.	Providing an overview and identifying gaps for further research.
Critical review	Provides a critical analysis of existing literature.	Synthesizing and reinterpreting existing knowledge.
Narrative review	Summarizes and discusses trends or themes in the literature.	Highlighting overarching themes and patterns.
Rapid review	Quickly assesses evidence in a constrained timeframe.	Supporting timely decision-making with a condensed analysis.

The Essential Components of Systematic Reviews

A high-quality systematic review adheres to rigorous methodologies to minimize bias and enhance validity. Essential components include the following:

- *Clearly Defined Objectives*: Objectives should articulate the specific research question, typically framed using the Population, Intervention, Comparison, and Outcome (PICO) model.
- *Explicit Eligibility Criteria*: Inclusion and exclusion criteria ensure the selection of relevant studies, mitigating selection bias.
- *Reproducible Methodology*: Transparent methodologies enable other researchers to replicate the review, enhancing credibility.
- *Comprehensive Literature Search*: Searches should encompass multiple databases, using precise keywords to capture the breadth of available evidence.
- *Quality Assessment*: Tools such as the Cochrane Risk of Bias tool or Newcastle-Ottawa Scale evaluate the methodological rigor of included studies.
- *Systematic Data Extraction*: Extracted data should include study characteristics, outcomes, and potential biases to provide a comprehensive overview.

Checklist for Conducting Literature Reviews

Identify Review Type: Determine the review type (e.g., systematic, integrative) based on your research goals.
Define Research Scope: Clearly state your objectives, research questions, and inclusion criteria.
Gather Relevant Literature: Use credible databases and ensure comprehensive coverage.
Analyze and Synthesize: Organize findings logically and provide critical insights.
Cite Properly: Use consistent and accurate citation styles.
Focus on Relevance: Include only studies that contribute meaningfully to your review objectives.
Provide Clear Structure: Use headings and subheadings to guide the reader.
Proofread: Check for clarity, coherence, and grammatical accuracy.

Evaluating Evidence Quality and Risk of Bias

Quality assessment ensures the reliability of conclusions drawn from SRs. Three primary dimensions are evaluated:

- *Internal Validity*:
  - Refers to the ability to avoid systematic errors through robust study designs.

  - Common biases include selection bias, performance bias, detection bias, and attrition bias.

- *External Validity*: Examines the generalizability of findings to broader populations.
- *Precision and Magnitude of Results*: Statistical tools assess the direction, magnitude, and confidence intervals (CI) of effects.

Assessments should ideally involve independent, blinded evaluators to reduce subjective bias. Utilizing standardized reporting guidelines such as CONSORT (for clinical trials), STROBE (for observational studies), and PRISMA (for systematic reviews and meta-analyses) enhances consistency and transparency.

Extraction, Analysis, and Interpretation of Results

The synthesis of data in systematic reviews involves summarizing findings and performing statistical analyses where applicable.

- *Data Synthesis*: Data from individual studies are compiled into summary tables, highlighting key characteristics and outcomes.
- *Statistical Methods*: Meta-analyses employ statistical tools to quantitatively aggregate results, generating overall effect sizes. A common graphical representation is the *forest plot*, which illustrates:
  - *Effect Size (e.g., Odds Ratio)*: Represented by squares, with larger squares indicating studies with greater weight.
  - *Confidence Intervals*: Horizontal lines depict the precision of estimates, with narrower intervals signifying higher precision.
- *Heterogeneity Assessment*:
  - Variability among study results is examined using statistical tests such as Cochran's Q test or the I^2 statistic.
  - High heterogeneity may necessitate subgroup analyses to identify factors contributing to variability.
- *Subgroup Analyses and Sensitivity Testing*:
  - Subgroup analyses evaluate outcomes within specific subsets of participants, while sensitivity tests determine the robustness of findings under varying assumptions.

Presentation of Results (Fig. 5.1 SR Flow Chart and Timelines)

Effective presentation of SR results involves both textual and visual elements, ensuring the findings are accessible to a professional audience.

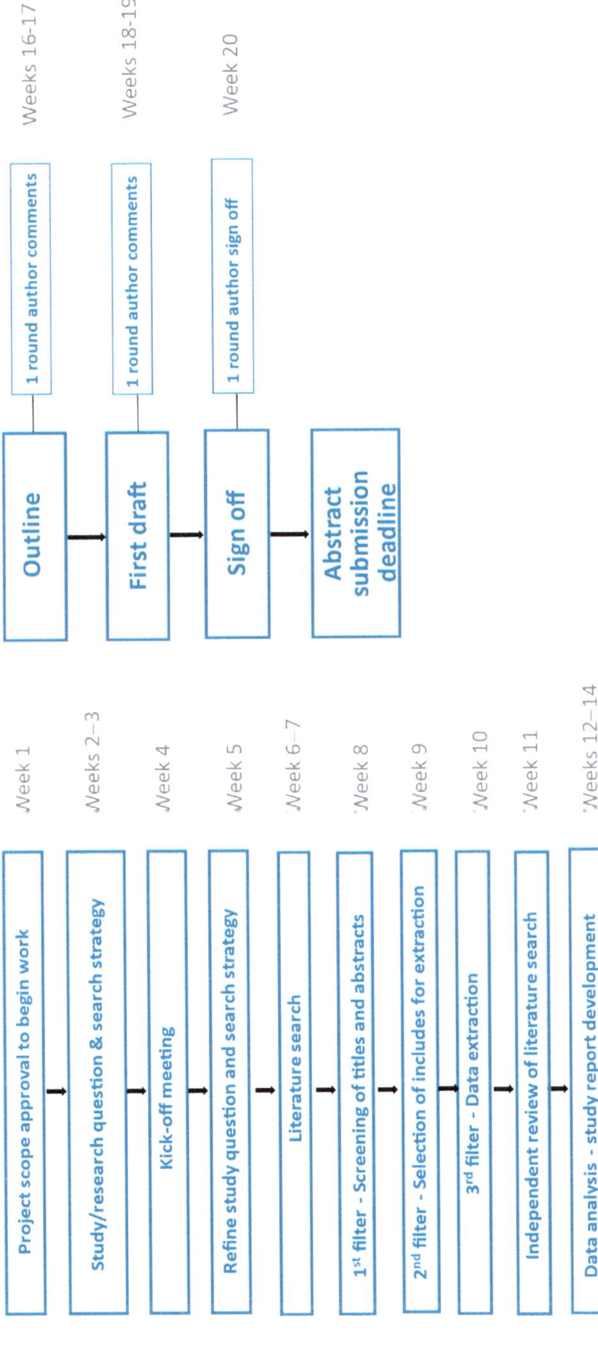

Fig. 5.1 Systematic review—Research, Report, and Abstract development flow chart

- *Textual Summaries*:
  - Results should be described in a structured narrative, highlighting significant findings and their implications.
  - Comparisons with prior research and discussion of clinical relevance strengthen the interpretative value.
- *Visual Representations*:
  - Graphs such as forest plots, funnel plots, and scatter plots effectively communicate statistical data.
  - Tables summarizing study characteristics, biases, and key outcomes provide a comprehensive overview.
- *Interpreting Findings*:
  - Emphasis should be placed on clinically significant outcomes rather than purely statistical significance.
  - Discussion sections should address the practical implications of findings, potential limitations, and directions for future research.

Challenges in Systematic Reviews

- *Publication Bias*:
  - The tendency to publish studies with positive results can skew meta-analytical outcomes.
  - Funnel plots and Egger's tests help detect asymmetry indicative of bias.
- *Heterogeneity*:
  - Divergence in study populations, interventions, and outcomes can complicate data synthesis.
  - Random-effects models accommodate heterogeneity but require cautious interpretation.
- *Data Quality and Availability*:
  - Poor-quality or incomplete data undermine the reliability of conclusions.
  - Sensitivity analyses and exclusion of low-quality studies can mitigate these issues.

SRs and meta-analyses are indispensable tools in advancing clinical research and informing evidence-based practice. Rigorous methodologies, coupled with transparent reporting and critical evaluation, are essential for deriving meaningful insights. As the volume of research grows, fostering proficiency in conducting and interpreting SRs will remain a priority for the scientific community.

Encouraging collaboration among researchers, clinicians, and statisticians will further enhance the quality and applicability of SRs. By adhering to best practices in analysis and presentation, SRs can continue to serve as a foundation for evidence-based decision-making, improving patient outcomes and advancing medical science.

Case Study: Company X's Scientific Communication Strategy for the Oncology Monoclonal Antibody (mAB)

Introduction

Company X, a global leader in oncology, has been at the forefront of developing and launching breakthrough cancer therapies. One such example is mAB, an anti-PD-L1 immunotherapy that has become a cornerstone treatment in several cancer indications, including non-small cell lung cancer (NSCLC), triple-negative breast cancer (TNBC), and bladder cancer. Since its approval, Company X has demonstrated an exemplary scientific communication strategy, incorporating a wide range of channels and approaches to engage HCPs, regulatory authorities, and patients. This case study illustrates how Company X implemented a comprehensive scientific communication plan, leveraging major congress presentations, a robust publication timeline, and innovative digital strategies to ensure widespread dissemination of mAB's clinical data and therapeutic benefits.

Publication Strategy and Timelines.

Company X's publication strategy for mAB was built around aligning the dissemination of clinical data with key regulatory milestones and pivotal oncology congresses. This strategy ensured that each phase of the drug's development was supported by transparent, high-impact scientific communication.

Timely Publication of Key Clinical Data.

Company X's scientific publication plan for mAB followed a well-structured timeline. The company strategically published results from major clinical trials such as IMpower110 and IMvigor130—landmark trials in lung and bladder cancer, respectively—in high-impact journals like *The New England Journal of Medicine* (NEJM) and *The Lancet Oncology*. These publications were synchronized with Company X's regulatory submissions to the FDA and EMA, ensuring that key trial results were available to support drug approval decisions.

For example, in the case of the IMpower150 study, which investigated mAB in combination with bevacizumab and chemotherapy for NSCLC, Company X published the primary results soon after obtaining FDA approval. The study demonstrated significant overall survival benefits, which was a critical milestone in shaping mAB's adoption in clinical practice. By following this publication strategy, Company X ensured that oncologists and HCPs had timely access to the data needed to make informed prescribing decisions, increasing the drug's market uptake.

Handling Complex Authorship.

A challenge often faced by large-scale trials like IMpower and IMvigor is managing the complexities of authorship. These trials often involve numerous investigators across multiple countries, each contributing to different aspects of the study.

Company X adhered to the ICMJE guidelines to ensure that all contributors were properly recognized, maintaining transparency and fairness in the authorship process. This approach helped prevent disputes and ensured the credibility of published findings, which is crucial for fostering trust within the scientific and medical communities.

Congress Presentations: Maximizing Engagement.

Presenting mAB's clinical data at major oncology congresses has been a cornerstone of Company X's scientific communication strategy. The American Society of Clinical Oncology (ASCO), the European Society for Medical Oncology (ESMO), and the World Conference on Lung Cancer (WCLC) are key venues where Company X has consistently showcased new data for mAB.

High-Impact Presentations at ASCO and ESMO.

In 2016, mAB's IMvigor211 trial results were presented at ASCO, highlighting its efficacy in advanced bladder cancer. The presentation generated significant interest from the oncology community, given the unmet need for effective treatments in bladder cancer. Company X's ability to showcase high-quality clinical data on such a prestigious stage demonstrated the value of mAB and cemented its place in oncological treatment paradigms. In fact, subsequent post-marketing data was further disseminated at ASCO and ESMO in the following years, continuously reinforcing mAB's long-term safety and efficacy profile.

These congress presentations were not only crucial for increasing HCP engagement but also strategically placed Company X in the spotlight as a leader in immuno-oncology. Company X's congress strategy was complemented by well-designed booths at these congresses, providing HCPs the opportunity to directly engage with Company X's medical team, review detailed study data, and gain a deeper understanding of mAB's clinical benefits.

Poster Presentations and Advisory Boards.

For data that was not yet ready for full publication, Company X utilized poster presentations at congresses to disseminate interim results and exploratory findings. Poster sessions offer an informal yet effective platform for engaging with the medical community, allowing for direct discussions between Company X's investigators and attending clinicians. This feedback was invaluable for refining research focus areas and understanding how the scientific community viewed mAB's evolving role in oncology treatment.

Additionally, Company X held advisory boards in conjunction with these congresses. These smaller, invite-only meetings allowed Company X to gather insights from key experts about mAB's positioning in clinical practice. Experts' feedback was instrumental in shaping subsequent studies and identifying new opportunities for expanding mAB's label.

Digital Communication Strategies: Virtual Congresses and Symposia.

The digital transformation of medical communication became more pronounced during the COVID-19 pandemic, and Company X was quick to adapt. The company leveraged virtual congresses, digital symposia, and webinars to continue engaging HCPs and disseminating mAB's data when in-person events were not possible.

During major oncology congresses, Company X hosted virtual symposia, allowing HCPs from around the world to access the latest mAB data without the need for travel. These symposia were designed to be interactive, offering live Q&A sessions with leading oncologists who had experience prescribing mAB. By enabling real-time discussions, Company X ensured that HCPs remained engaged and informed, even in a remote setting.

Company X also developed a series of on-demand webinars that focused on mAB's clinical applications across different tumor types. These webinars featured data reviews from the IMpower and IMvigor trials, discussions on RWE, and insights into patient selection for immunotherapy treatments. By offering this content online, Company X expanded its reach to HCPs who may not have been able to attend live congress sessions, ensuring the continued dissemination of critical scientific information.

Globalization of Scientific Communication: Navigating Regulatory and Regional Nuances.

mAB's global rollout required Company X to adopt a flexible but cohesive scientific communication strategy, as it was introduced in multiple regions with varying regulatory requirements and healthcare landscapes. Company X carefully tailored its communication strategies to fit the regulatory expectations of different markets, particularly the FDA in the United States, the EMA in Europe, and local health authorities in Asia and Latin America.

FDA and EMA Approvals: Synchronizing Communication Efforts

Company X ensured that its communication efforts surrounding mAB's FDA and EMA approvals were tightly aligned but also region-specific where necessary. For instance, when mAB was approved by the FDA for NSCLC based on the IMpower150 trial, Company X promptly published the clinical data in peer-reviewed journals and presented the findings at relevant US-based congresses. Simultaneously, Company X was preparing similar submissions to the EMA for approval in Europe, ensuring that data was available to European regulators and HCPs through timely publications and congress presentations.

Adapting to Localized Communication Needs.

In addition to harmonizing global communication efforts, Company X also tailored its messaging to meet the specific needs of local markets. For example, in Japan, Company X collaborated with local KOLs to ensure that the clinical benefits of mAB were communicated in a way that resonated with Japanese oncologists, who have a distinct approach to immunotherapy based on regional patient characteristics and treatment preferences.

Strengthening Ethical Perspectives: Transparency and Real-World Evidence.

Company X has emphasized ethical communication throughout mAB's lifecycle, with a focus on transparency, post-marketing surveillance, and real-world data reporting. The latter is increasingly important for assessing the long-term safety and effectiveness of oncology drugs in broader patient populations.

Company X published results from real-world studies, such as those evaluating mAB in patients with advanced NSCLC in routine clinical practice, through various channels, including congresses, peer-reviewed journals, and digital platforms. This commitment to continuously reporting both clinical trial and real-world data reinforces Company X's ethical responsibility to provide HCPs with the most comprehensive information on mAB's use in diverse patient populations.

Summary

Company X's scientific communication strategy for mAB exemplifies how a well-executed, multi-faceted approach can drive the successful adoption of a breakthrough oncology drug. By aligning publication timelines with key regulatory milestones, leveraging congress presentations and digital platforms to engage HCPs, and maintaining a strong ethical focus through the dissemination of RWE, Company X has ensured that mAB's clinical benefits are communicated clearly and effectively across the globe. This case study highlights how strategic scientific communication can support not only the success of a pharmaceutical product but also the broader goal of improving patient outcomes through transparency, education, and engagement with the medical community.

Chapter 6
Clinical Study

Key Highlights

- Clinical trials progress through phases I-IV to evaluate safety, efficacy, and long-term impact.
- Regulatory bodies (FDA, EMA) enforce compliance with GCP and ethical considerations.
- Decentralized clinical trials (DCTs) leverage digital health tools to enhance accessibility.
- Observational studies provide real-world insights to complement RCTs.
- Patient-centric approaches improve recruitment, adherence, and trial success.

Clinical trials represent the foundation of medical research, offering essential data that directs the development of innovative treatments and drugs. These studies are indispensable for determining the efficacy, safety, and appropriate usage of new medical interventions before their broader adoption in healthcare settings. In particular, pharmaceutical companies rely heavily on clinical trials to navigate the rigorous regulatory landscape and bring novel therapies to market. This chapter will provide a comprehensive overview of clinical trials, highlighting their definition, classical phases, regulatory frameworks, key participants, and critical documents. By delving into the intricate processes governing clinical research, this chapter will demonstrate the pivotal role clinical trials play in advancing medical science, especially within the pharmaceutical industry. The aim is to provide both an educational framework for those unfamiliar with clinical trials and a detailed exploration for professionals working in the field of drug development.

A. Krendyukov, *Medical Affairs' Fundamentals: The Next Generation*,
https://doi.org/10.1007/978-3-031-92588-7_6

Definition of Clinical Trials

Clinical trials are carefully designed research studies conducted on human volunteers to evaluate new medical, surgical, or behavioral interventions. These trials are the gold standard for determining whether new treatments, including drugs, medical devices, or therapeutic strategies, are safe and effective for human use. By applying controlled research methodologies, clinical trials aim to generate unbiased and reproducible results that can be used to inform healthcare decisions. The design and execution of clinical trials are strictly regulated to ensure the reliability of the data and the protection of trial participants. For pharmaceutical companies, clinical trials are an essential step in the drug development process, facilitating the transition from laboratory research to clinical practice.

Classical Phases of Clinical Trials (Fig. 6.1)

Clinical trials are typically conducted in distinct phases, each tailored to address specific questions about the safety, efficacy, and optimal use of the intervention.

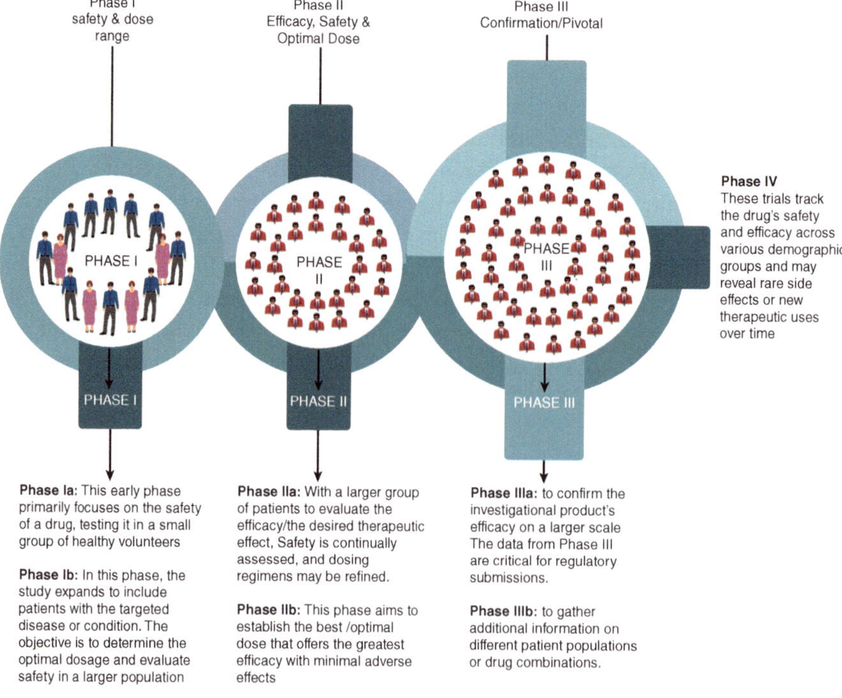

Fig. 6.1 Classical phases of clinical trials

Phase Ia and Ib (Safety and Dosage)

- *Phase Ia:* This early phase primarily focuses on the safety of a drug, testing it in a small group of healthy volunteers (20–80 people). The primary goal is to assess the drug's pharmacokinetics, including how it is absorbed, metabolized, and excreted. The data from this phase help identify the most frequent and serious adverse effects.
- *Phase Ib:* In this phase, the study expands to include patients with the targeted disease or condition. The objective is to determine the optimal dosage and evaluate safety in a larger population, while still closely monitoring adverse effects.

Phase IIa and IIb (Efficacy and Side Effects)

- *Phase IIa:* With a larger group of patients (100–300), this phase evaluates the efficacy of the drug, determining whether it produces the desired therapeutic effect. Safety is continually assessed, and dosing regimens may be refined.
- *Phase IIb:* This phase aims to establish the best dose with the least side effects. The drug is tested in a larger group of patients to identify the dose that offers the greatest efficacy with minimal adverse effects.

Phase IIIa and IIIb (Confirmation)

- *Phase IIIa:* Involving 1000–3000 patients, this phase seeks to confirm the drug's efficacy on a larger scale. The goal is to demonstrate that the drug is effective, monitor its side effects, and compare it with standard treatments. The data from Phase III are critical for regulatory submissions.
- *Phase IIIb:* Conducted after the submission of a New Drug Application (NDA) but before FDA or EMA approval, this phase may gather additional information on different patient populations, dosing strategies, or drug combinations.

Phase IV (Post-Market Surveillance) After regulatory approval, Phase IV trials monitor the drug's long-term effects in the general population. These trials track the drug's safety and efficacy across various demographic groups and may reveal rare side effects or new therapeutic uses over time.

Comparison of Randomized Controlled Trials and Observational Studies

Randomized Controlled Trials (RCTs) RCTs are considered the highest standard in clinical research. Participants are randomly assigned to treatment or control groups, reducing bias and providing reliable data for comparing the intervention with standard treatments or placebos. Randomization ensures that any differences in outcomes are due to the intervention rather than other factors.

Observational Studies In contrast, observational studies do not involve any intervention by the researcher. Instead, they monitor the effects of treatments, risk factors, or behaviors as they occur naturally. These studies are typically used for identifying trends or associations rather than establishing cause-and-effect relationships.

Key Differences

- *Methodology:* RCTs involve active intervention with random assignment, whereas observational studies passively observe real-world outcomes without manipulating the variables.
- *Outcomes:* RCTs aim to establish causality, while observational studies focus on identifying associations.
- *Reliability:* The controlled nature of RCTs typically provides stronger evidence due to their ability to eliminate confounding variables, and at the same time, narrow inclusion criteria can limit the extrapolation of the clinical outcome to border patient populations within intended indication and therapeutic areas.

Regulatory Framework of Clinical Trials

Clinical trials operate under a strict regulatory framework to ensure participant safety and data integrity.

Good Clinical Practice (GCP): GCP is an internationally recognized ethical and scientific standard for the design, conduct, and reporting of clinical trials. It ensures the credibility of the trial data and the protection of the participants' rights. GCP principles, rooted in the Declaration of Helsinki, are enforced by regulatory bodies worldwide, including the FDA and EMA.

International Council for Harmonisation (ICH): The ICH, comprising regulatory authorities and the pharmaceutical industry, provides guidelines to harmonize drug development processes across regions. ICH E6 (R2) offers a unified standard for conducting clinical trials, facilitating the mutual recognition of clinical data by regulatory agencies in different regions.

Declaration of Helsinki: First introduced in 1964, the Declaration of Helsinki sets ethical guidelines for medical research involving human subjects. It emphasizes informed consent, the protection of participant welfare, and the necessity of independent ethics committee review.

Role of Regulatory Bodies (FDA and EMA): The FDA in the United States and the EMA in the European Union are the primary authorities responsible for ensuring the safety and efficacy of new treatments. These agencies oversee clinical trial protocols, monitor compliance with GCP, and grant approval for the marketing of new drugs.

Key Actors in Clinical Trials

Clinical trials involve multiple stakeholders, each playing a crucial role in the successful execution of the study.

Investigators: Principal investigators lead the clinical trial, ensuring adherence to the protocol and regulatory requirements. They are responsible for patient recruitment, data collection, and ensuring participant safety throughout the study.

Sponsors: Sponsors, typically pharmaceutical companies, fund and manage the clinical trial. They are involved in designing the study, selecting investigators, providing the investigational product, and submitting trial data to regulatory authorities.

Regulatory Authorities: The FDA, EMA, and other regulatory bodies ensure compliance with clinical trial regulations. They review protocols, monitor trial conduct, and safeguard participant rights.

Ethics Committees/Institutional Review Boards (IRBs): These independent bodies review the ethical aspects of clinical trials to ensure that the rights and welfare of participants are protected.

Key Documents in Clinical Trials

Clinical Trial Protocol: The protocol outlines the trial's objectives, design, methodology, and statistical considerations, serving as a guide for investigators and a reference for IRBs/ethics committees.

Informed Consent Form (ICF): This document explains the potential risks, benefits, and alternatives of the trial to participants, ensuring that they fully understand the study before agreeing to participate.

Investigator's Brochure (IB): This comprehensive document compiles all preclinical and clinical data on the investigational product, providing essential information for investigators.

Case Report Forms (CRFs): CRFs are used to capture participant data during the trial, ensuring accurate and consistent reporting across all sites.

Regulatory Submission Documents: These include the necessary documents for submitting trial data to regulatory agencies for approval, reporting progress, and presenting final results.

Final Report (Clinical Study Report/CSR): At the end of the clinical trial, a final report summarizes the methodologies, findings, and conclusions. This document is vital for interpreting and publishing the trial's results.

The Evolving Landscape of Clinical Trials in Pharmaceutical Development

The field of clinical trials is undergoing significant transformation, driven by technological advancements, evolving regulatory frameworks, and an increasing focus on patient-centric approaches. For pharmaceutical companies, these changes are redefining the way clinical studies are conducted, providing new opportunities to enhance efficiency, reduce costs, and improve patient outcomes. This chapter explores the current trends shaping the clinical trial landscape, highlighting the rise of DCTs, the integration of digital health technologies, patient-centric approaches, regulatory adaptations, globalization, and the growing emphasis on sustainability. By understanding these emerging trends and challenges, sponsors, researchers, and regulatory bodies can better navigate the complexities of modern clinical research.

The Rise of Decentralized Clinical Trials (DCTs)

DCTs represent a fundamental shift in how clinical studies are designed and conducted. Unlike traditional site-based trials, DCTs leverage digital technologies, such as telemedicine, wearables, and remote monitoring, to conduct trials outside of centralized research centers. This model allows patients to participate from their homes or local healthcare facilities, eliminating geographic barriers and reducing the burden of travel. The COVID-19 pandemic significantly accelerated the adoption of DCTs, demonstrating their feasibility and effectiveness in maintaining clinical trial continuity during times of restricted in-person interactions. For pharmaceutical companies, DCTs offer the potential to streamline trial logistics, increase participant diversity, and accelerate the drug development process.

Advantages of DCTs

- *Enhanced Participant Access and Convenience:* DCTs enable wider geographic reach, allowing more diverse patient populations to participate in clinical trials. This is particularly beneficial for rare disease studies, where recruiting sufficient patient numbers at centralized sites can be challenging.
- *Improved Data Collection:* Continuous monitoring through digital tools, such as wearables, provides real-time data, potentially increasing the accuracy and granularity of the collected information. This can lead to more robust safety and efficacy assessments, as data are captured directly from patients in their real-world environments.

Challenges in Implementing DCTs

- *Ensuring Data Quality and Integrity:* While DCTs offer the advantage of remote data collection, ensuring the accuracy and reliability of data remains a challenge.

Variability in how patients use remote devices and self-report information may introduce inconsistencies in the dataset.
- *Regulatory and Ethical Considerations:* Adapting existing regulations to accommodate DCT methodologies requires careful consideration. Regulatory agencies, such as the FDA and EMA, are working to update guidelines to ensure that remote trials maintain high standards of patient safety, data privacy, and ethical conduct.

Role of Digital Health Technologies and Technology Integration in Clinical Trials

Digital health technologies, including wearables, mobile apps, and electronic health records (EHRs), are increasingly becoming integral to clinical trial designs. These technologies not only enhance data collection but also improve patient engagement by providing real-time feedback on their health status.

- *Wearables and Mobile Apps:* Devices like smartwatches and health-monitoring apps track vital signs, physical activity, and patient-reported outcomes. For example, heart rate, glucose levels, and sleep patterns can be continuously monitored, providing a more comprehensive picture of a drug's impact on a patient's daily life.
- *Artificial Intelligence (AI) and Machine Learning (ML):* AI and ML are being used to optimize trial design, predict patient outcomes, and identify the most suitable candidates for trials through sophisticated algorithms. These technologies improve efficiency in data analysis, allowing researchers to identify patterns and trends more quickly than with traditional methods.

Challenges in Data Management and Cybersecurity

As clinical trials become increasingly digital, managing large volumes of data from various sources presents significant challenges. Ensuring data integrity, particularly when integrating information from wearables, apps, and EHRs, requires robust data governance frameworks. Additionally, the shift toward digital data collection necessitates heightened cybersecurity measures to protect sensitive patient information from breaches and unauthorized access.

Shift Toward Patient-Centric Approaches in Clinical Trials

The focus on patient centricity in clinical trials has gained momentum as sponsors and researchers recognize the value of involving patients in the trial design and execution process. Patient-centric trials aim to prioritize the needs, preferences, and experiences of participants, improving recruitment, retention, and overall trial success.

- *Incorporating Patient Feedback:* By integrating patient feedback into the design of clinical trials, sponsors can select endpoints that are more meaningful to patients, potentially increasing adherence and engagement. For example, selecting endpoints that reflect improvements in quality of life rather than solely clinical metrics can make trials more appealing to participants.
- *Communication Strategies:* Clear and effective communication with patients is critical for maintaining engagement. Patient-centric trials use lay language in consent forms and patient-facing documents to ensure participants fully understand the risks and benefits of the study.

Challenges in Patient Recruitment and Retention

- *Diverse Recruitment:* Recruiting diverse patient populations remains a major challenge for pharmaceutical companies conducting clinical trials. Barriers such as socioeconomic status, geographic location, and language can limit participation. Addressing these barriers through DCTs and tailored recruitment strategies can enhance the generalizability of trial results.
- *Retention Strategies:* Ensuring patient retention throughout the trial requires thoughtful strategies to reduce dropout rates. These may include offering flexible visit schedules, reducing travel burdens through home visits or virtual consultations, and providing ongoing support to patients throughout the trial.

Regulatory Challenges and Adaptations

Regulatory bodies are evolving their frameworks to keep pace with technological advancements and innovative trial designs. The rise of DCTs, AI-driven data analysis, and patient-centric approaches necessitates updated guidelines that balance innovation with patient safety.

- *Regulatory Flexibility:* Regulatory agencies such as the FDA and EMA have shown an increasing willingness to accommodate new methodologies, providing more flexible pathways for trial approval. For instance, adaptive trial designs, which allow for modifications based on interim results, have become more widely accepted, offering a more efficient approach to drug development.
- *Ethical Considerations:* As trials become more complex, navigating ethical concerns, particularly in areas like gene editing and personalized medicine, remains

crucial. Regulators are tasked with ensuring that trial designs respect patient autonomy while minimizing risks.

Balancing Innovation and Patient Safety

While innovation in clinical trial design offers numerous benefits, maintaining stringent safety standards is paramount. Pharmaceutical companies must strike a delicate balance between embracing new technologies and methodologies and ensuring that the safety of trial participants is never compromised. This is particularly important in trials involving cutting-edge treatments, such as gene therapies and personalized medicine, where the long-term effects may be uncertain.

Globalization: Trends in Conducting Multi-Regional Clinical Trials

As pharmaceutical companies seek to expedite drug development and approval, multi-regional clinical trials (MRCTs) have become increasingly common. These trials, conducted across multiple countries and regions, allow researchers to assess the safety and efficacy of drugs in diverse populations, providing a more comprehensive understanding of a drug's global applicability.

- *Leveraging Global Data:* By conducting trials in different regions, sponsors can generate data that reflect varying genetic, environmental, and cultural factors. This global approach to data collection helps to accelerate regulatory approvals by providing a broader evidence base for safety and efficacy claims.

Challenges in Regulatory Compliance and Cultural Considerations

- *Navigating Regulatory Requirements:* Conducting MRCTs requires careful navigation of different regulatory environments. While harmonization efforts, such as ICH guidelines, have made this easier, varying local requirements can still pose challenges for sponsors.
- *Ethical and Cultural Differences:* Cultural considerations, such as differing attitudes toward consent or medical interventions, can impact trial design and conduct. Sponsors must adapt their approach to informed consent and patient engagement to respect local customs and regulatory requirements.

Sustainability in Clinical Trials

In response to growing awareness of environmental issues, the pharmaceutical industry is increasingly adopting sustainable practices in clinical trial design and execution. This includes reducing the carbon footprint of trials, minimizing waste, and sourcing materials sustainably.

- *Green Protocols:* Initiatives such as paperless trials, reduced travel through DCTs, and the use of electronic data capture systems are helping to minimize the environmental impact of clinical trials. Sponsors are also exploring ways to reduce the use of disposable materials and ensure the ethical sourcing of trial supplies.

Challenges in Implementing Green Protocols

- *Logistical and Financial Barriers:* While sustainability is a priority, implementing green protocols can pose logistical and financial challenges. For example, transitioning to electronic systems may require significant upfront investment, and reducing travel can complicate trial logistics in regions where digital infrastructure is lacking.

Summary

The landscape of clinical trials is evolving rapidly, driven by technological innovations, patient-centric approaches, globalization, and sustainability initiatives. These changes offer pharmaceutical companies new opportunities to enhance the efficiency, diversity, and ethical standards of clinical research, but they also present significant challenges. As the industry moves forward, maintaining a balance between innovation, patient safety, regulatory compliance, and environmental responsibility will be critical. The future of clinical trials is poised to be more inclusive, efficient, and responsive to the needs of both patients and society at large, ultimately contributing to the faster and safer development of life-saving treatments.

Expanding the Horizon of Clinical Trials: Case Studies, Regulatory Innovations, and Future Trends

The landscape of clinical trials is shifting rapidly, driven by technological innovations, regulatory adaptations, and emerging ethical challenges. This section delves into real-world case studies, explores global regulatory perspectives, and examines the future of clinical trials. By understanding the successes and challenges in DCTs, the ethical considerations of AI-driven trials, and the global regulatory response, we can better navigate the evolving complexities in clinical research.

Real-World Case Studies: Decentralized Clinical Trials and AI Integration

Case Study: Company Y's Decentralized Approach During the COVID-19 Pandemic

The COVID-19 pandemic provided a catalyst for the rapid implementation of DCTs. A leading example is Company Y's phase III clinical trial for its COVID-19 vaccine, which adopted decentralized elements to maintain momentum despite restrictions on in-person interactions. With the support of remote monitoring technologies and digital communication platforms, Company Y recruited a diverse global patient population and conducted essential trial operations, including data collection and patient monitoring, remotely.

The use of telemedicine, eConsent forms, and wearable devices allowed real-time data collection from participants across various regions, without compromising the trial's quality or regulatory compliance. The success of Company Y's DCT highlights how this type of clinical trial can not only reduce patient burden by eliminating travel but also increase recruitment and retention rates by offering more flexibility for participants.

Case Study: AI in Recruitment for Rare Disease Trials

The integration of AI has revolutionized clinical trial recruitment, particularly for rare diseases where identifying eligible patients can be time-consuming and costly. The pharmaceutical company Y used AI to screen potential candidates for a rare disease trial by analyzing EHRs and genomic data to identify individuals who matched the complex eligibility criteria. This reduced the recruitment phase by 30%, significantly cutting costs and accelerating the overall timeline of the trial.

AI's ability to parse large datasets, identify patterns, and predict patient responses not only streamlines the recruitment process but also enhances the precision of data analysis, allowing for more efficient and targeted clinical trials, especially in complex areas like oncology and personalized medicine.

Expanding Regulatory Perspectives: Global Adaptations and Innovations

Regulatory Innovation in China and Japan

As the global clinical trial landscape evolves, countries like China and Japan are adapting their regulatory frameworks to accommodate innovations such as DCTs and AI-driven analytics. In China, regulatory authorities have adopted a more flexible approach, issuing guidelines that support the use of digital technologies in trials while also focusing on fast-tracking drug approvals for critical therapies. China's

National Medical Products Administration (NMPA) has adopted a regulatory sandbox approach, where novel methodologies can be tested under a more lenient but controlled environment. This allows pharmaceutical companies to conduct trials more efficiently while still adhering to safety and ethical standards.

Similarly, Japan's Pharmaceuticals and Medical Devices Agency (PMDA) has introduced adaptive trial designs to accommodate new technologies, such as AI for predictive analytics and personalized treatments. These adaptive designs allow mid-trial modifications based on interim results, making trials more responsive to early findings and potentially reducing overall costs and timeframes.

Cultural and Societal Considerations in Global Trials

Conducting multi-regional clinical trials presents unique challenges, particularly when considering cultural and societal differences. Informed consent, for example, is perceived differently across regions, with varying degrees of trust in healthcare authorities and differences in how risks are communicated and understood.

In Asia, where hierarchical structures and collective decision-making may be more prevalent, sponsors must adapt their trial designs to ensure patients fully understand the implications of participation. Furthermore, local regulatory bodies may impose additional layers of oversight to ensure ethical standards are met, making early engagement with these authorities essential for trial success. Pharmaceutical companies must also navigate differing societal norms, such as varying levels of acceptance toward genetic testing or gene therapies, which may impact recruitment and retention strategies.

Ethical Considerations in Modern Clinical Trials

AI-Driven Trials: Addressing Algorithmic Bias

While AI offers enormous potential in streamlining clinical trials, its use also presents ethical challenges. AI systems are trained on historical data, which may carry biases—such as underrepresentation of minority groups or specific patient demographics—that could lead to skewed outcomes. For example, if an AI algorithm used in patient recruitment primarily reflects data from Caucasian populations, it may underperform when applied to more diverse populations, thus impacting the trial's inclusiveness and generalizability.

To mitigate this, regulatory bodies and sponsors must ensure that AI systems are trained on diverse and representative datasets. Ethical frameworks for AI-driven trials should include continuous monitoring to detect potential biases and ensure the fairness of AI-driven decision-making.

Gene Editing and Personalized Medicine: Long-Term Ethical Implications

Clinical trials involving gene editing, such as CRISPR-based therapies, present complex ethical issues, particularly around long-term safety and informed consent. The irreversible nature of gene therapies raises concerns about unforeseen side effects that may only manifest years after the trial concludes, posing challenges for both trial sponsors and regulatory bodies.

Informed consent is crucial but complicated, as participants must fully understand the potential for long-term risks. Moreover, there is the issue of equitable access—gene therapies are often expensive and may only be accessible to certain socioeconomic groups, raising concerns about inequality in healthcare access. Long-term monitoring, post-trial surveillance, and ethical oversight must be part of the regulatory framework to ensure that these therapies are not only safe but also equitably distributed.

Enhancing Participant Engagement in the Digital Era

Telemedicine and Continuous Patient Feedback Loops

Telemedicine has emerged as a key tool in DCTs, allowing continuous engagement with participants through remote consultations and monitoring. By using telemedicine platforms, researchers can maintain frequent touchpoints with patients, providing real-time feedback on health metrics and addressing concerns or side effects as they arise.

Incorporating patient feedback into trial designs—both during and after the trial—helps ensure that the trial remains responsive to participant needs. Feedback loops, facilitated through mobile apps or online portals, allow patients to share their experiences, contributing to patient-centered outcomes and potentially improving retention rates.

Empowering Patients through Digital Platforms

Digital platforms, including patient advocacy groups and social media channels, are empowering patients to take a more active role in clinical research. These platforms not only offer information about ongoing clinical trials but also serve as forums for patient discussions, creating communities where participants can share experiences and encourage others to enroll in trials. Pharmaceutical companies are increasingly leveraging these platforms to reach diverse populations and to better communicate the benefits of trial participation, ultimately improving recruitment and retention.

Long-Term Data Collection and Monitoring in DCTs

Wearables for Chronic Condition Monitoring

The long-term use of wearables for chronic disease monitoring, such as for cardiovascular disease or diabetes, has proven invaluable in providing continuous real-world data. However, this presents unique challenges in terms of data integrity and patient compliance. Devices may malfunction, or patients may fail to wear them consistently, leading to gaps in data collection. Ensuring proper device calibration, technical support, and patient training is critical to maintaining high data quality.

Patient Privacy in Long-Term Digital Data Collection

With the increased use of digital health technologies, patient privacy concerns have become paramount. The continuous collection of sensitive health data, such as glucose levels or heart rate, over extended periods requires robust cybersecurity protocols to prevent data breaches. Sponsors must ensure that data collected from wearables or mobile devices is encrypted, securely stored, and accessible only to authorized personnel, while remaining compliant with regional privacy regulations like the General Data Protection Regulation (GDPR) in Europe or the Health Insurance Portability and Accountability Act (HIPAA) in the United States.

Future-Focused Technologies: Blockchain, Virtual Trials, and Genomic Data

Blockchain in Clinical Trials

Blockchain technology has the potential to revolutionize the transparency and security of clinical trials. By creating an immutable ledger of all trial data and processes, blockchain can ensure the integrity of the data, prevent tampering, and enable secure sharing of information between stakeholders. For DCTs, blockchain could streamline the consent process, track data provenance, and ensure compliance with ethical and regulatory standards.

Virtual Trials and the Metaverse

The use of virtual and augmented reality in clinical trials—commonly referred to as "metaverse trials"—offers an immersive way for participants to engage with trial protocols remotely. Virtual trials could allow participants to interact with HCPs and trial data in a virtual environment, reducing the need for physical site visits and improving patient convenience.

CRISPR and Genomic Data Integration

The integration of genomic data into clinical trials, particularly those involving CRISPR-based gene editing, presents opportunities to develop highly personalized treatments. However, this also raises ethical and regulatory challenges, particularly around data ownership and long-term monitoring of genetic modifications. Ensuring that genomic data is handled with care and protected from misuse is crucial as these technologies become more widespread.

Conclusion

As clinical trials continue to evolve, stakeholders—including pharmaceutical companies, regulators, and researchers—must collaborate to embrace innovation while upholding ethical and safety standards. Emerging technologies like AI, blockchain, and wearables offer immense potential to improve trial efficiency, but their implementation must be carefully monitored to ensure fairness and data integrity. By prioritizing patient-centric approaches, addressing global regulatory challenges, and adopting sustainable practices, the future of clinical trials can be more inclusive, transparent, and effective. It is now the responsibility of the industry to take these advancements forward, ensuring that clinical trials remain at the forefront of medical innovation while always prioritizing the safety and well-being of participants.

Case Study: Adaptive Design in Oncology Trials: Transforming the Paradigm of Clinical Research (based on A. Krendyukov and M. Lerchenmueller "Challenges and opportunities to the implementation of adaptive design in Phase III oncology trials: results from a cross-sectional analysis" ESMO 2023)

Background

Cancer continues to impose a significant burden globally, with Europe alone witnessing 2.7 million new cases and 1.27 million cancer-related deaths in 2022. Globally, cancer medicine spending reached $223 billion in 2023, and projections estimate an increase to $409 billion by 2028. Amid this landscape, Innovative Medicinal Products (InMPs) are imperative to address the high unmet medical needs in oncology. However, the path to developing these therapies is fraught with challenges, including high R&D costs averaging over $4 billion per approved medication, of which more than 50% stems from clinical research.

Oncology trials, representing a substantial portion of all clinical trials, frequently employ novel trial designs to optimize efficiency and accelerate the development timeline. Among these, adaptive clinical trials have emerged as a transformative alternative to traditional RCTs with conventional designs. Adaptive trials, which allow for prospectively planned modifications based on accumulating data, represent a pivotal advancement in addressing the complexities of oncology research. This section explores the benefits of adaptive design, its advantages over conventional methods, and its potential in shaping the future of oncology clinical trials.

Understanding Adaptive Design

Adaptive clinical trials introduce flexibility into the trial design, enabling real-time adjustments based on pre-specified rules and interim analyses (Fig. 6.2). These adjustments can include changes to sample size, patient population, treatment arms,

Adaptive Trial Types (examples)	Description
Group sequential design	Prospective interim analysis with pre-specified criteria for terminating the trial
Sample size	Prospectively planned modifications to the sample size based on interim estimates (e.g. unblinded sample size adaptation/re-estimation)
Patient population (e.g. adaptive enrichment)	Adaptive modification of the patient population based on comparative interim results
Treatment arm selection (pick-the-winner/drop-the-loser; adaptive dose ranging, etc.)	Adding or terminating arms (dropping the inferior treatment group(s), modifying treatment arms and/or adding additional arms; to allocate more patients to treatment doses of interest, reducing allocation of patients to doses that appear non informative, etc.)
Patient allocation (randomization or treatment switching, etc.)	Adaptation based either on comparative baseline characteristics or based on comparative outcome data
Endpoint selection	Adaptive modification to the choice of primary endpoints based on comparative interim results
Seamless Phase II/III	Allows combined objectives (Phase II and III) moving from Phase II (investigational stage) to Phase III (efficacy or confirmatory) without stopping the patient enrolment process
Biomarker adaptive	Allows adaptations based on interim analysis of the treatment responses of biomarkers (can be used to select patient populations for subsequent trials, identify the natural course of a disease)
Multiple design features	Combination of two or more adaptive features (complex)
Special considerations and topics	• Simulation in adaptive trial planning • Bayesian design • Time-to event setting • Potential surrogate or intermediate endpoints • Secondary endpoints and safety considerations • Design changes/hypothesis change

Fig. 6.2 A number of different types of adaptive trial design with specific considerations have been developed and described in guidelines issued by competent authorities over the last decade. (Adapted from FDA guidelines (listed as one of the references))

or endpoints. Regulatory bodies, including the FDA, have issued guidelines categorizing various adaptive trial designs, such as:

- Group Sequential Design: Incorporating interim analyses for early trial termination or continuation.
- Sample Size Reassessment: Adjusting sample sizes based on interim data.
- Treatment Arm Selection: Implementing "pick-the-winner" or "drop-the-loser" methodologies.
- Seamless Phase II/III Designs: Combining phases to streamline development.

This flexibility not only enhances efficiency but also allows researchers to explore broader scientific questions and adapt to evolving clinical insights.

Comparative Analysis: Adaptive vs. Conventional Design

In a cross-sectional analysis of Phase III trials conducted for oncology InMPs approved by the US FDA in 2018, adaptive designs accounted for 25% of trials, maintaining parity with data from 2013. Among adaptive trials, 81% focused on solid tumors, highlighting their potential value in this therapeutic area. Key differences between adaptive and conventional trials elucidate their operational advantages and challenges:

- *Trial Duration:*
  - Adaptive trials exhibited a median duration of 43 months, significantly shorter than the 60 months observed in conventional trials ($p = 0.0129$).
  - This temporal compression translates into substantial cost savings, estimated at 30% (approximately $1 billion for the 14 approved oncology products in 2018).
- *Site and Patient Distribution:*
  - Adaptive trials involved a higher median number of sites (203) compared to conventional trials (106).
  - Patient enrollment per site was significantly lower in adaptive trials (3.8 patients) than in conventional trials (35.9 patients), suggesting more dispersed patient recruitment and stricter monitoring.
- *Endpoints:*
  - Both adaptive and conventional trials demonstrated comparable primary endpoints (median 1.6 for adaptive vs. 1.5 for conventional).
  - However, adaptive trials had a higher median number of secondary endpoints (18 vs. 12), underscoring their capacity for a more comprehensive assessment.

These distinctions reveal adaptive design's ability to streamline trials while generating richer, multidimensional data, crucial for advancing oncology research.

Benefits of Adaptive Design

- Efficiency and Speed:
  - By reducing trial duration, adaptive designs expedite the availability of life-saving therapies to patients.
  - The integration of multiple phases (e.g., seamless Phase II/III) minimizes redundancy and accelerates development timelines.
- Cost Savings: The shortened duration and streamlined processes in adaptive trials contribute to significant cost reductions, mitigating the financial burden of oncology trials.
- Flexibility and Responsiveness: Real-time modifications based on interim data allow for optimized trial conduct, enhancing decision-making and resource allocation.
- Enhanced Scientific Insights: The broader scope of endpoints and adaptive methodologies enrich the depth of clinical and scientific insights, fostering innovation.

Challenges and Limitations

Despite its advantages, adaptive design poses certain challenges that warrant careful consideration:

- Operational Complexity: The need for real-time data monitoring, interim analyses, and protocol adjustments introduces logistical and operational challenges.
- Regulatory Scrutiny: Regulatory authorities require detailed pre-specifications of adaptation rules, demanding rigorous planning and documentation.
- Transparency: Limited details on adaptive methodologies and interim decision-making processes highlight the need for greater transparency to bolster credibility.
- Resource Intensiveness: Higher site participation and intensive monitoring increase operational demands compared to conventional designs.

Practical Implications and Recommendations

To maximize the benefits of adaptive design in oncology trials, stakeholders must address these challenges and refine implementation strategies:

- Regulatory Engagement: Early and continuous collaboration with regulatory agencies ensures compliance and alignment with methodological expectations.
- Technological Integration: Leveraging advanced data management and analytical tools facilitates real-time monitoring and adaptation.
- Enhanced Transparency: Providing detailed methodologies and decision-making processes fosters trust among stakeholders and regulators.
- Training and Expertise: Building capabilities in adaptive trial design among research teams enhances execution and efficacy.
- Collaboration and Data Sharing: Collaborative efforts, such as those outlined in the European Commission's Pharmaceutical Strategy for Europe, support the adoption of innovative trial designs and facilitate knowledge exchange.

Adaptive clinical trials represent a paradigm shift in oncology research, offering significant advantages over conventional designs. Their ability to reduce trial durations, optimize resource allocation, and generate multifaceted insights positions them as a cornerstone for advancing cancer therapeutics. While challenges related to operational complexity and transparency persist, targeted efforts to address these issues can unlock the full potential of adaptive design.

As cancer incidence and associated costs continue to rise, the imperative for innovative and efficient research methodologies becomes ever more critical. By embracing adaptive trial designs, the pharmaceutical industry can accelerate the development of transformative therapies, addressing the urgent needs of oncology patients and reshaping the future of clinical research.

Biostatistics in Clinical Research

Biostatistics plays a critical role in clinical research, serving as the backbone for data analysis, interpretation, and ultimately, decision-making in the field of medical science. It provides the tools needed to derive meaningful inferences from data collected during clinical trials, ensuring that medical interventions are both safe and effective. Without robust statistical methods, clinical data could easily lead to misleading conclusions, jeopardizing patient safety and delaying the approval of potentially life-saving treatments. Regulatory bodies such as the FDA and EMA rely heavily on statistical evidence to assess the efficacy and safety of new pharmaceutical products. Therefore, a firm understanding of statistical principles and their correct application is essential for scientists, clinicians, and reviewers involved in medical research.

However, despite the critical nature of biostatistics, statistical errors remain prevalent in clinical trial design, data analysis, and manuscript review. These errors can stem from flawed experimental design, incorrect statistical analyses, or misinterpretation of results. The consequences of these mistakes are significant, leading to overestimated treatment effects, misidentification of drug efficacy, and wasted resources in both time and funding. Addressing these issues is essential to enhance the reproducibility and credibility of scientific findings.

This chapter aims to explore ten common statistical mistakes in biostatistics, providing insight into how these errors occur and offering guidance on how to avoid them. By understanding and mitigating these statistical pitfalls, researchers can ensure the validity and reliability of their clinical findings.

Common Statistical Pitfalls in Clinical Research

Absence of an Adequate Control Condition or Group

A common statistical error in clinical research is the failure to incorporate an adequate control group. In clinical trials, outcome measures are often assessed at multiple time points to determine the effects of an intervention. However, changes in outcomes may arise not only from the intervention itself but also from external factors such as time, placebo effects, or natural recovery. Without a control group, it is impossible to distinguish between changes caused by the intervention and those due to other variables.

For instance, if a study only measures pre- and post-intervention outcomes without a control group receiving a placebo or no treatment, the researchers cannot claim that the observed effects are due to the treatment alone. Furthermore, poorly designed control groups—such as those that differ in baseline characteristics or have inadequate sample sizes—can introduce bias and lead to incorrect conclusions.

To avoid this, clinical trials must include well-designed control groups that are comparable to the experimental group in all aspects except the intervention under investigation. Randomization and blinding are also essential components in reducing bias and ensuring the reliability of study results.

Interpreting Comparisons Between Two Effects Without Direct Comparison

Another frequent mistake is comparing two separate effects without directly testing for differences between them. Researchers often fall into this trap when they observe that one group shows a statistically significant result while another does not and then infer that the effect size is larger in the former group.

For example, in a clinical trial, Group A might show a statistically significant reduction in symptoms after treatment, while Group B does not. It may be tempting to conclude that the treatment was more effective for Group A than for Group B. However, this inference is incorrect unless a direct statistical comparison between the two groups is conducted. Such errors arise from treating two independent statistical results as if they were dependent, when in fact, only a formal statistical test comparing the two groups can provide the correct inference.

To resolve this issue, researchers should use appropriate statistical tests, such as interaction terms in analysis of variance (ANOVA) or direct comparisons of effect sizes between groups using statistical tests like unpaired t-tests.

Inflating the Units of Analysis

In clinical research, the unit of analysis should reflect the number of independent observations that can be freely varied, which often represents the number of participants in a study. However, researchers sometimes make the mistake of inflating the number of observations by including repeated measures (e.g., multiple time points from the same participants) as independent data points.

For instance, if a study with 10 participants reports the results of pre- and post-intervention measures, treating these 20 observations as independent inflates the degrees of freedom, resulting in inflated significance values. This leads to overconfidence in the observed effects, which may not truly exist when the dependency between repeated measures is considered.

The solution to this issue lies in using mixed-effects models, which account for both within-subject and between-subject variability. These models allow researchers to include all available data without violating the assumption of independence and are particularly useful in longitudinal studies where multiple measurements are taken from the same individuals over time.

Spurious Correlations Due to Outliers or Subgroups

Spurious correlations arise when statistical tests are influenced by outliers or hidden subgroups within the data. In clinical trials, an outlier can artificially inflate or deflate the correlation coefficient between variables, leading to incorrect conclusions about the strength of associations.

Similarly, if data from two distinct subgroups are pooled together without acknowledging the differences between them, spurious correlations may emerge. This is particularly problematic when the subgroups differ systematically in key variables but are analyzed as a single group.

To avoid spurious correlations, researchers should carefully examine their data for outliers and consider using robust statistical techniques such as bootstrapping or winsorizing to minimize the impact of extreme values. Additionally, identifying and accounting for subgroups within the dataset is essential before performing correlation analyses.

Use of Small Sample Sizes

Small sample sizes are a persistent problem in clinical research, especially in studies involving rare diseases or small patient populations. While small samples may be unavoidable in some cases, they reduce the statistical power of a study, increasing the likelihood of Type I and Type II errors. This means that studies with small sample sizes are more prone to both false positives (i.e., identifying an effect when none exists) and false negatives (i.e., missing an effect that is actually present).

The significance of a small sample size is compounded by the fact that small samples are more likely to deviate from normality, which can skew results in parametric analyses. Moreover, small sample sizes lead to inflated effect sizes, making it more likely that a study will overestimate the true effect of an intervention.

To mitigate this risk, researchers should aim to conduct power analyses prior to the study to ensure that the sample size is sufficient to detect meaningful effects. In cases where large sample sizes are not feasible, researchers should acknowledge the limitations of their study and avoid making overly strong conclusions based on small datasets.

Statistical Power

Statistical power, the probability that a study will detect an effect if it exists, is intrinsically linked to the size of the sample. A study with low power runs a high risk of both Type I errors (false positives) and Type II errors (false negatives). This balance is essential for the correct interpretation of clinical data.

When statistical power is low, the study becomes less sensitive to detecting true effects, and the margin for error increases. Even when a significant result is observed, small sample sizes often inflate the effect size, providing an inaccurate representation of the true relationship between variables. Furthermore, the variability in small sample sizes can lead to skewed results, particularly if the sample does not adequately represent the population of interest. For instance, in genetic studies or precision medicine, underpowered studies can lead to overestimation of the effect size of a treatment, leading to unrealistic expectations about therapeutic efficacy.

It is crucial to conduct an a priori power analysis during the design phase of a clinical study. This analysis estimates the required sample size based on anticipated effect sizes, the desired significance level (alpha), and the acceptable probability of Type II errors (beta). In situations where recruiting large samples is not feasible, particularly for rare conditions, researchers should consider alternative designs, such as crossover designs or Bayesian analyses, which are better suited for small sample scenarios. Bayesian methods, which will be discussed below, provide a flexible framework for drawing conclusions when data is sparse.

Circular Analysis (Double Dipping)

Circular analysis, or "double dipping," occurs when researchers use the same data to both define and test a hypothesis. This introduces bias and inflates the likelihood of finding a significant result, even when no true effect exists. Circular analysis is often seen in studies where researchers retrospectively select features of the data (e.g., splitting the dataset based on post hoc criteria) and then use these features to generate inferences.

An example of this might be a study in which researchers initially fail to find a significant effect but then identify a subset of the data that shows a significant result after subgrouping. This result is often a statistical artifact of the data selection process rather than a genuine effect.

To avoid circular analysis, researchers should predefine their hypotheses and analysis plans before examining the data. If post hoc analyses are necessary, they should be clearly labeled as exploratory and not used to make definitive claims about the effects under investigation.

Flexibility of Analysis: p-Hacking

P-hacking refers to the practice of manipulating data analysis procedures to achieve statistically significant results. This can include selectively reporting results, changing outcome measures after data collection, or conducting multiple analyses and only reporting the ones that yield significant results. P-hacking increases the likelihood of Type I errors, leading to false positives that are not representative of true effects.

To combat p-hacking, researchers should preregister their study protocols and analysis plans, ensuring that the methods are predetermined and not altered based on the data. Transparency in reporting is also essential, with researchers clearly distinguishing between confirmatory and exploratory analyses.

Failing to Correct for Multiple Comparisons

When a study involves multiple statistical comparisons, the probability of obtaining a significant result by chance increases. If this is not accounted for through corrections such as the Bonferroni correction or false discovery rate adjustments, the study's results are likely to include false positives.

For example, in neuroimaging studies, thousands of voxels may be tested for statistical significance. Without correcting for the sheer number of comparisons, significant results may appear simply due to random noise rather than any true effect.

Researchers should always apply appropriate corrections when conducting multiple comparisons, particularly in exploratory analyses with large datasets. While these corrections reduce the risk of false positives, they also increase the risk of false negatives, so researchers must balance caution with practicality.

Over-interpreting Non-significant Results

Non-significant results in clinical trials are often misinterpreted as evidence that no effect exists. However, a non-significant p-value does not distinguish between the absence of an effect and an underpowered study that failed to detect an existing effect. Misinterpretation of non-significant results can lead to the incorrect conclusion that a treatment is ineffective when, in fact, the study was simply not sensitive enough to detect a small effect.

Researchers should be cautious when interpreting non-significant results and should report confidence intervals and effect sizes alongside p-values to provide a more nuanced understanding of the data. Bayesian statistics or equivalence testing can also be used to assess whether a non-significant result truly supports the null hypothesis.

Confusing Correlation with Causation

Perhaps the most well-known statistical error is the confusion between correlation and causation. A significant correlation between two variables does not imply that one causes the other. This error is especially common in observational studies, where researchers may be tempted to draw causal inferences from correlational data.

To avoid this pitfall, researchers should rely on well-designed RCTs to establish causal relationships. In cases where RCTs are not feasible, alternative statistical methods, such as mediation analysis or hierarchical modeling, can help elucidate potential causal mechanisms.

From Correlation to Causation: A Subtle but Important Distinction

Statistical relationships, particularly correlations, are often misinterpreted in clinical studies, leading to overreach in the conclusions drawn. Correlation, by definition, only implies that two variables change in relation to one another; it does not imply that one causes the other. This misunderstanding can result in misguided clinical recommendations, potentially harming patients if a supposed cause-effect relationship is later debunked. The temptation to infer causality is particularly strong when correlations align with existing hypotheses or theoretical expectations, but this leap from association to causality must be approached with caution.

The issue of mistaking correlation for causation is particularly prevalent in observational studies, where randomization is not feasible, and confounding variables may influence both the exposure and the outcome. For example, in epidemiological studies examining lifestyle factors, it may be tempting to claim that an increase in physical activity directly causes a reduction in cardiovascular risk. However, this association could be driven by other factors such as diet, genetics, or healthcare access, none of which are adequately controlled for in the analysis. A well-conducted RCT, on the other hand, can provide the robust evidence necessary to establish causality, as it minimizes bias through random assignment of participants.

Bayesian Approaches: Enhancing Clinical Data Interpretation

Bayesian statistics offer a powerful alternative to traditional frequentist methods, particularly in the context of clinical trials where sample sizes are limited or prior knowledge about the intervention exists. While frequentist approaches rely on the concept of long-run frequencies to make inferences about data, Bayesian methods incorporate prior probabilities and update these beliefs as new data are gathered. This allows for a more nuanced interpretation of clinical trial data, especially when dealing with uncertainty or variability in results.

In a Bayesian framework, the probability of an effect is directly computed, integrating prior knowledge (e.g., previous trials, expert opinion) with observed data to produce a posterior probability distribution. For instance, in early-phase clinical trials, where data is scarce but there may be prior evidence from preclinical studies or similar interventions, Bayesian methods allow researchers to make informed decisions about the likely efficacy of a treatment. This contrasts with the traditional reliance on p-values, which do not provide direct information about the probability of an effect.

One significant advantage of Bayesian methods is their flexibility in interpreting non-significant results. In frequentist analysis, a non-significant p-value is often interpreted as evidence that the null hypothesis is true, though this is not technically accurate. In Bayesian analysis, however, non-significant results can be combined with prior information to provide a more informative conclusion. Researchers can use Bayesian models to estimate the likelihood that a treatment is effective, even if the current data does not reach conventional significance thresholds.

For example, consider a small trial investigating a new cancer therapy. A Bayesian approach could incorporate prior data from related therapies to estimate the probability that the new treatment is beneficial, providing a clearer picture of its potential value, despite the limited data from the current trial. By integrating prior knowledge with new evidence, Bayesian methods can lead to more robust decision-making in clinical trials, particularly when navigating the challenges of small sample sizes or ambiguous results.

Enhancing Statistical Rigor in Clinical Research

Biostatistics is an indispensable tool in clinical research, offering the methods and rigor needed to derive meaningful conclusions from data. However, the misuse of statistical techniques can lead to erroneous conclusions, which can have serious consequences for drug development, patient safety, and healthcare outcomes. By understanding and addressing common statistical mistakes, researchers can improve the reliability and reproducibility of clinical studies, ultimately contributing to more robust scientific knowledge and better healthcare interventions.

By integrating best practices in experimental design, analysis, and reporting, the scientific community can work towards minimizing these errors and fostering greater transparency and integrity in clinical research.

The importance of applying correct statistical principles in clinical research cannot be overstated. Statistical mistakes, whether arising from poor design, incorrect analyses, or misinterpretation of results, can significantly undermine the validity of clinical studies. More importantly, such errors can delay the development of new therapies, misinform clinical decision-making, and ultimately affect patient outcomes. This chapter has highlighted ten common statistical mistakes, providing practical solutions to address these issues.

To promote better scientific practice, it is crucial for researchers to adopt rigorous statistical methods, including the use of appropriate control groups, power calculations, mixed-effects models, and Bayesian statistics where applicable. Furthermore, transparency in reporting statistical methods, such as the pre-registration of studies and distinguishing between exploratory and confirmatory analyses, is essential to foster reproducibility and trust in scientific findings.

As the field of clinical research continues to evolve, with increasingly complex data and innovative methodologies, staying vigilant against statistical errors is more important than ever. By adhering to sound statistical principles, researchers can contribute to a body of evidence that is not only more reliable but also more directly translatable into clinical practice, ultimately benefiting patients and advancing medical knowledge.

Chapter 7
Education and Training in Pharmaceutical Companies with a Focus on Medical Affairs

Key Highlights

- Continuous learning ensures compliance with global regulatory standards (FDA, EMA, ICH) and ethical marketing practices.
- Training programs focus on emerging therapeutic areas, including gene therapy, immuno-oncology, and biologics.
- Digital transformation has led to the adoption of webinars, e-learning modules, virtual reality (VR), and gamification.
- Real-world evidence (RWE) and patient-centric approaches are integral to modern medical education.
- Cross-functional collaboration enhances the role of Medical Affairs in engaging with regulatory, clinical, and commercial teams.

In a dynamic healthcare environment, medical professionals must stay current with the latest research, drug developments, and regulations. The importance of continuous education for Medical Affairs is particularly crucial as they interact with healthcare providers (HCPs), regulators, and internal stakeholders.

Medical Affairs professionals need to be well-versed in medical science, regulatory changes, and the latest therapeutic developments to provide value to the healthcare ecosystem and improve patient outcomes.

Ensuring Compliance and Regulatory Adherence

One of the primary drivers of training programs is ensuring that all employees, particularly Medical Affairs professionals, remain compliant with global regulations set by authorities like the FDA, EMA, and ICH.

Pharmacovigilance Training: Regular training ensures proper reporting of adverse events and that safety monitoring protocols are followed meticulously.

A. Krendyukov, *Medical Affairs' Fundamentals: The Next Generation*,
https://doi.org/10.1007/978-3-031-92588-7_7

Ethical Marketing Practices: Medical Affairs professionals are responsible for ensuring that all interactions with HCPs are ethical and adhere to regulatory guidelines, minimizing the risk of violations.

Enhancing Knowledge of Emerging Therapeutics

The rapid advancement of therapeutic areas such as gene therapy, immuno-oncology, and biologics necessitates continuous education. Medical Affairs teams must be well-versed in the mechanisms of action, clinical data, and patient outcomes to engage meaningfully with HCPs.

Developing Knowledge of Complex Therapies: Through training programs, Medical Affairs professionals stay updated on emerging treatments and best practices, enabling them to effectively disseminate this knowledge to external stakeholders.

Building Competency in Medical Affairs

Medical Affairs professionals require a broad set of competencies that go beyond scientific knowledge. Communication, regulatory understanding, and business acumen are essential.

Cross-Functional Skills: Medical Affairs teams must bridge internal departments such as Regulatory Affairs, Clinical Operations, and Commercial teams. Training helps develop skills that allow them to collaborate across functions effectively.

Improving Patient Outcomes and Public Health

A well-trained Medical Affairs team ensures that accurate, up-to-date medical information reaches HCPs, which can directly impact treatment decisions and improve patient outcomes. Training programs emphasize patient-centric approaches to enhance healthcare delivery.

Evolving Dynamics in Medical Education

Over the last few decades, the nature of medical education has transformed, shaped by technology, evolving stakeholder needs, and time constraints faced by professionals.

New Stakeholders, Including Patients: The modern patient is more informed, often seeking medical information through online platforms. Medical Affairs professionals are increasingly interacting with patients and patient advocacy groups, requiring training in effective communication.

Empowering Patients: Medical Affairs professionals are now responsible for delivering information that helps patients make informed decisions, especially in areas like rare diseases and oncology.

The Need for More Creativity and Digital Innovation: The traditional classroom-based training model is being replaced by creative digital platforms. Modern medical education relies on engaging, interactive content that caters to different learning styles.

Engaging Content: The use of webinars, podcasts, and mobile applications is growing, offering flexible and creative ways to deliver educational content.

Addressing the Challenge of Time Constraints: Healthcare professionals are pressed for time, which has led to a shift towards modular, on-demand training formats.

Asynchronous Learning Tools: Short, focused learning modules allow medical professionals to fit education into their schedules without disrupting their daily responsibilities.

Fast-Paced Evolution of Medical Domains: The rapid development of new therapeutic areas and health trends, such as the COVID-19 pandemic, has demonstrated the need for agile learning mechanisms that provide real-time knowledge dissemination.

Real-Time Learning Tools: Virtual conferences and digital congress reports have become critical tools in keeping Medical Affairs teams updated on the latest research.

The Shift to Digital and Interactive Learning

Pharmaceutical companies are leveraging modern digital tools to make training accessible, scalable, and engaging. Interactive content encourages deeper learning and can be customized to individual needs.

Webinars and Virtual Training Sessions

Webinars offer live training opportunities where professionals can learn from experts in real-time, participate in Q&A sessions, and access panel discussions.

Benefits: On-demand replay options provide flexibility for those unable to attend live sessions.

Congress Reports and Real-Time Educational Content

Real-time updates from major medical congresses, such as ESMO and ASCO, provide Medical Affairs teams with the latest information on drug developments, clinical trials, and therapeutic advancements.

Continuous Learning: Companies disseminate key congress takeaways to Medical Affairs teams, enabling them to stay informed.

Online Educational Materials

Self-paced online courses, e-learning modules, and educational portals offer opportunities for continuous learning.

Key Platforms: Examples include platforms such as ESMO, ASCO, and Medscape, which offer specialized modules on therapeutic areas.

Collaboration with CME Providers and Medical Societies

The Role of CME in Professional Development

Accredited Continuing Medical Education (CME) programs provide unbiased, up-to-date education for medical professionals. Pharma companies collaborate with CME providers to offer certified educational content that meets industry standards.

Partnerships with Medical Societies

Medical societies play a crucial role in establishing high standards for education. Pharma companies partner with societies like the American Medical Association (AMA) to develop educational modules that meet the evolving needs of the medical community.

Ensuring Scientific Integrity and Ethical Boundaries

It is essential that educational content be based on science, free of commercial bias, and prioritize patient welfare. Partnerships between pharma companies, CME providers, and medical societies ensure that the educational content remains credible.

Interactive Training Modules and Programs

Interactive training is revolutionizing education, offering engaging, adaptable formats that improve retention and applicability.

Case-Based Learning Modules: Simulated real-world medical scenarios help professionals apply theoretical knowledge to practical situations, improving clinical decision-making skills.

Interactive Simulations and Virtual Reality (VR): The use of AI-driven simulations and VR creates immersive training environments that mimic real-world settings.

Enhanced Learning: Professionals can practice handling complex cases, new medical devices, or clinical procedures in a safe, controlled environment.

Gamification of Learning: Gamification uses elements of game design (e.g., rewards, leaderboards) to engage and motivate learners. It has proven effective in increasing engagement and retention.

Tracking and Assessment Tools: Data analytics and tracking tools provide valuable insights into learning progress. Companies can assess the efficacy of their training programs and provide personalized learning paths.

Challenges and Opportunities in Implementing Education and Training

Challenges

Resistance to Change: Some employees may resist transitioning from traditional learning methods to modern digital formats.

Balancing Global vs. Localized Needs: Companies must develop training programs that cater to both global standards and local regulations.

Ensuring Compliance: Regulatory requirements necessitate the highest educational standards, requiring continuous updates and audits of training content.

Opportunities

Scalable, Global Training Programs: Digital tools enable companies to roll out standardized training programs across regions.

Collaborative Partnerships: Industry-academia collaborations offer innovative solutions and evidence-based training.

Personalized Learning: AI-powered platforms are creating customized learning experiences, adapting content based on learner progress.

Reinforcing the Importance of Education and Training in Medical Affairs

Continuous education is crucial for the success of pharmaceutical companies, particularly in Medical Affairs, where the role requires up-to-date scientific, regulatory, and communication skills.

The future of training in pharma lies in innovative digital tools, interdisciplinary collaboration, and continuous, patient-centered learning.

Encouraging Ongoing Commitment to Education

Pharma companies must invest in innovative and continuous education initiatives that prioritize patient care, collaboration, and scientific integrity.

The Role of Collaboration

By working together with CME providers, medical societies, and academic institutions, pharmaceutical companies can ensure high educational standards and foster a culture of continuous learning.

Patient-Centric Education

Training programs should ultimately focus on improving patient outcomes, fostering trust, and ensuring that the medical community is equipped with the knowledge to provide the highest standard of care.

Enhancing Medical Affairs Education Through Real-World Examples, Digital Tools, and Interdisciplinary Collaboration

Case Study: Adaptation to Digital Tools During the COVID-19 Pandemic
The COVID-19 pandemic accelerated the transition to digital platforms for education and training. Pharmaceutical companies were forced to rethink their traditional, in-person training programs. One company, in particular, adapted by developing a robust, fully digital education program for their Medical Affairs team, utilizing a combination of webinars, VR simulations, and on-demand online modules.

The use of webinars allowed the company to deliver real-time training to its global team, despite travel restrictions and physical distancing measures. Expert-led sessions were hosted virtually, followed by Q&A panels and interactive discussions. These webinars were recorded and made available on-demand, enabling Medical Affairs professionals to revisit the material at their convenience.

The company also integrated VR technology into its training for medical device simulations. The VR modules provided a hands-on, immersive learning experience where professionals could simulate medical procedures, interact with new devices, and respond to clinical scenarios in a virtual setting. This led to an enhanced understanding of the product's practical use, which is crucial for Medical Affairs teams as they educate HCPs.

The feedback from this digital adaptation was overwhelmingly positive. The team reported an increase in engagement with the content, and post-training assessments showed improved retention of key information. The flexibility offered by these digital tools also catered to the professionals' busy schedules, allowing them to balance their educational needs with daily responsibilities.

The Role of Data Analytics in Training

Personalized Learning Paths

As training programs evolve, many pharmaceutical companies are leveraging data analytics to create personalized learning paths tailored to the needs of their Medical Affairs professionals. These adaptive learning platforms analyze user data, such as progress, strengths, and weaknesses, to customize educational content and deliver more focused, effective learning experiences.

For example, a pharmaceutical company implemented a learning management system (LMS) that monitored user progress through various training modules. The system could identify which concepts Medical Affairs professionals struggled with, such as specific scientific knowledge or communication techniques. Based on these insights, the system recommended additional learning modules, providing targeted support to address gaps in knowledge. This personalized approach ensures that learners focus on the areas where they need the most improvement, enhancing the overall effectiveness of the training.

Learning Analytics for Impact Measurement

Data analytics also enable companies to measure the impact of their training programs. By tracking completion rates, quiz scores, time spent on modules, and performance improvements post-training, companies can assess the efficacy of their educational initiatives.

For instance, after implementing a new oncology training program, a pharmaceutical company used analytics to measure the program's success. Metrics such as the number of participants who completed the training, assessment scores before and after the program, and follow-up surveys with HCPs indicated that the training resulted in a 30% increase in knowledge retention among Medical Affairs professionals. These insights were crucial in refining future training programs, ensuring continuous improvement and alignment with the company's strategic goals.

Continuous Learning Opportunities

The pharmaceutical industry is constantly evolving, and as such, Medical Affairs professionals need to engage in continuous learning to stay updated on the latest therapeutic developments, medical devices, and regulatory changes. Pharmaceutical companies are increasingly promoting lifelong learning by offering ongoing education and training programs that align with the latest industry standards.

For example, a global pharmaceutical company designed an ongoing education initiative that allowed Medical Affairs professionals to participate in quarterly updates on emerging therapies in their therapeutic areas. This initiative included webinars, access to congress reports, and collaboration with external experts who shared insights on future trends in medicine. The program's continuous nature ensured that the team remained well-versed in cutting-edge science, which translated into better interactions with HCPs.

To add credibility to their roles, many Medical Affairs professionals seek certification in specialized areas, such as pharmacovigilance, medical writing, or regulatory affairs. Companies are increasingly collaborating with institutions to offer these certification programs, recognizing the value of formal accreditation in enhancing the credibility and skill sets of their teams.

One pharmaceutical company worked with a leading academic institution to offer a certification program in pharmacovigilance. The program provided Medical Affairs professionals with a thorough understanding of drug safety reporting, risk management, and compliance with regulatory standards. Upon completing the certification, team members reported a heightened confidence in their ability to address safety concerns and communicate effectively with regulatory authorities. This certification also enhanced the company's reputation, signalling its commitment to maintaining the highest standards in pharmacovigilance.

Interdisciplinary Training Programs

Medical Affairs professionals often collaborate with colleagues from Regulatory Affairs, Clinical Operations, R&D, and Commercial teams. To facilitate seamless collaboration, interdisciplinary training programs are becoming increasingly important. These programs ensure that Medical Affairs professionals have a well-rounded understanding of the company's scientific, regulatory, and business objectives.

An example of successful interdisciplinary training comes from a pharmaceutical company that developed a joint training initiative involving Medical Affairs, Regulatory Affairs, and R&D teams. The training focused on the development and launch of a new immunotherapy drug. Each department contributed expertise—Medical Affairs brought in insights on HCP engagement, R&D covered scientific and clinical trial data, and Regulatory Affairs ensured that the messaging complied with global regulatory requirements. This collaboration led to a unified strategy for the drug's launch and resulted in consistent, compliant communication across all touchpoints with HCPs and regulatory bodies.

Joint training sessions with other departments, such as Marketing or Compliance, enhance overall knowledge and improve operational efficiency. A pharmaceutical company, for example, organized a joint workshop involving Medical Affairs, Marketing, and Compliance teams to improve understanding of ethical promotion and off-label communication. By including Compliance professionals in the training, the Medical Affairs and Marketing teams gained a clearer understanding of the boundaries around promotional activities, ensuring that all future interactions were both ethical and compliant.

Overcoming Digital and Remote Learning Challenges

Although digital learning offers flexibility and scalability, it also presents certain challenges. Technology adoption, time management, and engagement issues can hinder the effectiveness of online learning programs. To overcome these barriers, companies must carefully consider the needs of their learners and implement strategies to foster engagement.

One common challenge is the lack of engagement in virtual environments. To address this, a pharmaceutical company incorporated interactive elements into their online modules, such as quizzes, live Q&A sessions, and discussion boards. These interactive features encouraged active participation, making the learning experience more dynamic.

Blended Learning Models

Blended learning, which combines online modules with in-person sessions, has emerged as a solution to many of the challenges associated with remote learning. This model offers the flexibility of digital learning while maintaining the personal interaction that comes with face-to-face training.

For example, a pharmaceutical company introduced a blended learning program for a product launch. The initial phase consisted of online, self-paced modules to familiarize Medical Affairs professionals with the new drug's clinical data. This was followed by in-person workshops where professionals could ask questions, discuss challenges, and participate in hands-on training sessions. This combination proved effective, as it allowed professionals to learn at their own pace while benefiting from the interactive elements of in-person training.

Soft Skills Development in Medical Affairs

Medical Affairs professionals need a unique blend of technical expertise and soft skills to thrive in their roles. Communication, relationship-building with healthcare professionals, problem-solving, and empathy are critical competencies that complement their scientific knowledge.

A pharmaceutical company recognized the need for soft skills training and implemented a program that focused on improving communication and relationship-building. The program included role-playing exercises where professionals practiced explaining complex scientific data to HCPs in a clear and empathetic manner. This helped Medical Affairs professionals build stronger connections with HCPs, ultimately leading to more meaningful, productive conversations.

As Medical Affairs professionals advance in their careers, they often transition into leadership roles. Leadership and management training therefore become essential in developing the skills necessary to lead teams, manage projects, and negotiate effectively.

One company introduced a leadership training program for senior Medical Affairs professionals that included modules on project management, strategic decision-making, and negotiation tactics. The program helped participants develop the skills needed to lead cross-functional teams, make strategic decisions, and negotiate with stakeholders effectively.

Knowledge Sharing Platforms and Future-Proofing Education in Medical Affairs

In the rapidly evolving pharmaceutical landscape, Medical Affairs teams are at the heart of the interaction between HCPs, regulatory bodies, and internal company stakeholders. To remain effective in this role, Medical Affairs professionals must have access to the latest medical knowledge, regulatory updates, and therapeutic advancements. Knowledge-sharing platforms—both internal and external—have become essential tools in this endeavor, allowing professionals to collaborate, exchange information, and continually develop their expertise.

Internal Knowledge Sharing Platforms

Pharmaceutical companies are increasingly recognizing the importance of *internal knowledge-sharing platforms*—or "learning communities"—to foster a culture of continuous learning and collaboration among Medical Affairs professionals. These platforms offer a structured yet flexible environment where professionals can exchange insights, share case studies, discuss challenges, and disseminate key information across teams in real time.

The Importance of Internal Knowledge Platforms

- *Real-Time Knowledge Dissemination*: Internal platforms enable the rapid dissemination of critical knowledge across the organization. Medical Affairs professionals frequently encounter new data from clinical trials, product updates, and regulatory changes. An internal platform allows this information to be shared instantly, ensuring that all team members are aligned with the latest developments.
- *Peer-to-Peer Learning*: One of the most powerful aspects of internal knowledge-sharing platforms is the peer-to-peer learning environment they foster. Medical Affairs professionals can share their experiences, offer insights on complex cases, and collaborate on problem-solving. This collaborative model encourages an ongoing exchange of ideas, which helps build a knowledge-rich organization.
- *Breaking Down Silos*: In large pharmaceutical companies, departments often work in silos, making cross-functional communication difficult. Internal platforms bridge these gaps, allowing for more fluid communication between Medical Affairs, Regulatory, R&D, and Commercial teams. As a result, knowledge sharing becomes a catalyst for innovation and efficiency across the entire organization.

- *Examples of Internal Knowledge Platforms*: One major pharmaceutical company implemented an internal knowledge-sharing platform designed to streamline access to scientific publications, clinical trial data, and regulatory updates. The platform features discussion boards, document libraries, and live Q&A sessions where experts within the company can field questions from Medical Affairs professionals. This structure ensures that all staff have the opportunity to contribute and benefit from the collective intelligence of the organization.

The Role of Technology in Internal Knowledge Sharing

Technology plays a crucial role in ensuring the effectiveness of these platforms. Many pharmaceutical companies have adopted cloud-based learning management systems (LMS) that support real-time updates, interactive features (such as webinars and live discussions), and personalized content feeds. These platforms integrate with mobile apps, allowing Medical Affairs professionals to access critical information anytime and anywhere. Additionally, data analytics built into these systems provide insights into usage patterns, helping companies refine the platform's offerings and improve user engagement.

External Knowledge Hubs

While internal platforms are critical, *external knowledge hubs* offer an equally important opportunity for Medical Affairs professionals to connect with broader networks of experts, thought leaders, and peers in the medical and scientific community. External platforms provide access to the latest industry-wide developments, helping professionals stay current on therapeutic advances and regulatory trends.

Engaging with Professional Medical Societies

Many external knowledge-sharing platforms are hosted by *professional medical societies*, such as the European Society for Medical Oncology (ESMO), the American Society of Clinical Oncology (ASCO), and the American Medical Association (AMA). These platforms offer valuable resources, including the following:

- *Access to Cutting-Edge Research*: Medical Affairs professionals can access the latest clinical research, conference reports, and peer-reviewed studies. Staying informed about these developments is crucial for engaging meaningfully with HCPs and regulatory bodies.

- *Collaborative Research Initiatives*: Some external hubs allow pharmaceutical companies to collaborate on research initiatives, offering opportunities to co-author studies or contribute to white papers. Engaging in these initiatives enhances a company's thought leadership position and ensures their Medical Affairs teams are part of shaping the future of medicine.

Open-Source Knowledge Platforms

In addition to professional societies, open-source platforms such as *PubMed* and *ResearchGate* offer access to a wide range of medical publications. Pharmaceutical companies can encourage their Medical Affairs teams to actively participate in these forums, whether by sharing their own research or contributing to ongoing discussions. This engagement strengthens the team's visibility and credibility within the global medical community.

Case Study: Leveraging External Knowledge Hubs
A leading global pharmaceutical company encouraged its Medical Affairs team to engage with external knowledge hubs like *ASCO's Virtual Meeting Library*, which provides on-demand access to presentations and discussions from oncology conferences. By participating in these virtual meetings, the Medical Affairs team gained insights into the latest trends in oncology research, which directly informed their engagement with HCPs. The result was an increase in HCP satisfaction, as the team was able to provide real-time, relevant information on new treatments and therapies.

Future-Proofing Education in Medical Affairs

As the pharmaceutical industry continues to evolve, Medical Affairs professionals must stay ahead of emerging trends and technologies that are transforming both healthcare and the practice of Medical Affairs. Future-proofing education ensures that professionals are equipped with the knowledge and skills needed to navigate this rapidly changing landscape.

Focus on Future Technologies in Learning

Technological advancements are playing a pivotal role in shaping the future of education for Medical Affairs professionals. Beyond adaptive learning and AI, several emerging technologies hold the potential to revolutionize the way professionals are trained.

Augmented Reality (AR) and Virtual Reality (VR)

Augmented reality (AR) and virtual reality (VR) are already being used to simulate complex medical procedures and enhance experiential learning. Medical Affairs professionals can use AR and VR to immerse themselves in clinical environments, allowing them to:

- *Practice Complex Procedures*: VR simulations provide a hands-on learning experience for Medical Affairs professionals who need to understand new medical devices, treatment procedures, or drug delivery methods.
- *Train on Patient Engagement*: AR can be used to simulate interactions with virtual patients, allowing professionals to practice explaining medical information, managing difficult conversations, or presenting clinical trial data.

Blockchain for Credentialing

Blockchain technology has the potential to *revolutionize credentialing* for medical education programs by providing a secure, tamper-proof system for tracking certifications and CME credits. Blockchain can store and verify credentials in a transparent, decentralized way, ensuring that Medical Affairs professionals' qualifications are easily verifiable by employers and regulatory authorities.

- *Transparency and Accountability*: By using blockchain, professionals can have a digital record of their certifications and completed training modules. This ensures that credentials are trustworthy and verifiable across organizations and regions.
- *Example in Practice*: Pharmaceutical companies can partner with academic institutions and certification providers to establish blockchain-based credentialing systems, allowing professionals to maintain a lifelong learning record that is recognized worldwide.

Emerging Skills and Trends in Pharmaceutical Education

Medical Affairs professionals must be prepared for several key trends that are reshaping the pharmaceutical landscape. As healthcare becomes more personalized and data-driven, new skills will be required.

Personalized Medicine and Gene Therapies

As *personalized medicine* and *gene therapies* become more prevalent, Medical Affairs professionals will need to deepen their understanding of genetics, biomarkers, and individualized treatment plans. Training programs must evolve to include these cutting-edge topics, preparing professionals to engage in complex discussions with HCPs about tailored treatment approaches.

Real-World Evidence (RWE)

With the increasing use of RWE to support regulatory approvals and post-market safety monitoring, Medical Affairs teams must be equipped with the skills to interpret and apply RWE. Training programs will need to cover topics such as data analytics, RWE methodologies, and the integration of RWE into regulatory and clinical decision-making processes.

Ethical Use of AI in Healthcare

As AI becomes more integrated into healthcare, Medical Affairs professionals will need to understand the *ethical implications* of using AI-driven tools. Training will need to cover responsible AI usage, addressing concerns such as data privacy, bias, and patient safety.

Encourage Peer-Led Training and Mentoring Programs

In addition to formal education and training modules, *peer-led training* and *mentoring programs* have emerged as invaluable tools for fostering professional growth and development within pharmaceutical companies. These programs facilitate knowledge transfer, promote experiential learning, and help build a collaborative learning culture.

Mentoring as a Key Educational Tool

Mentoring allows experienced professionals to share their expertise with junior Medical Affairs staff, offering guidance and support in navigating complex challenges. Mentorship programs are particularly effective for:

- *Career Development*: Mentors can help mentees navigate their career paths, offering advice on gaining new skills, building professional networks, and advancing within the organization.
- *Skill Building*: Mentorship provides an opportunity for hands-on learning, with mentors offering real-time feedback on areas such as communication, regulatory knowledge, and clinical decision-making.

Case Study: Examples of Mentorship Programs

One pharmaceutical company implemented a formal mentorship program aimed at *onboarding new Medical Affairs professionals*. Each new hire was paired with a senior mentor from within the company. These mentors provided guidance on the company's therapeutic areas, facilitated introductions to key internal and external stakeholders, and helped their mentees navigate the complexities of interacting with HCPs. The result was a smoother onboarding process, with new hires achieving competency faster and feeling more supported in their roles.

Another company established a *peer-led training group* focused on navigating ethical challenges in off-label communication. Senior Medical Affairs professionals led discussions and case studies with their peers, offering practical insights on how to manage regulatory compliance while effectively supporting HCPs. This peer-led approach helped create an open environment for sharing best practices and addressing common challenges.

Conclusion

Incorporating knowledge-sharing platforms, embracing future technologies, and fostering peer-led training programs are essential steps in future-proofing education for Medical Affairs professionals. By integrating internal and external knowledge hubs, companies can ensure that their teams remain at the forefront of medical advancements. Emerging technologies such as AR, VR, and blockchain offer exciting possibilities for enhancing the learning experience, while peer-led mentoring programs provide valuable opportunities for hands-on, experiential learning. Together, these strategies ensure that pharmaceutical companies continue to develop highly skilled Medical Affairs teams capable of navigating the evolving healthcare landscape.

Chapter 8
Engagement and Collaboration with External Experts and Medical Societies

Key Highlights

- External experts contribute to clinical research, regulatory approvals, and post-marketing surveillance.
- Collaboration with medical societies influences clinical guidelines, education, and healthcare policy.
- Regulatory compliance (EFPIA, PhRMA, Sunshine Act) ensures ethical and transparent expert interactions.
- Digital tools (AI-driven expert mapping, virtual advisory boards) enhance engagement and data-driven decision-making.
- Long-term expert relationship management fosters scientific credibility and advances medical innovation.

External experts, including key opinion leaders (KOLs), key external experts (KEEs), thought leaders (TLs), and subject matter experts (SMEs), play a crucial role in the pharmaceutical industry. These individuals have deep expertise in specific therapeutic areas, backed by their contributions to research, academia, and clinical practice. External experts provide valuable input across the drug development lifecycle, from preclinical research to post-marketing surveillance.

In the pharmaceutical landscape, the involvement of external experts is not only pivotal for scientific advancement but also for regulatory success. Their participation ensures that drug development aligns with the highest standards of clinical science and meets the needs of healthcare professionals and patients. Moreover, pharmaceutical companies rely on external experts to communicate complex scientific data to medical communities, enhancing the credibility of their therapeutic innovations.

Engaging with these experts requires a balance of strategic planning, compliance with ethical guidelines, and a focus on long-term relationship development. Regulatory frameworks, such as the European Federation of Pharmaceutical

A. Krendyukov, *Medical Affairs' Fundamentals: The Next Generation*,
https://doi.org/10.1007/978-3-031-92588-7_8

Industries and Associations (EFPIA) and Pharmaceutical Research and Manufactures of America (PhRMA) codes, outline specific criteria and transparency measures for interactions with external experts, ensuring that collaborations are both ethical and compliant with industry standards.

Role of Medical Societies: Medical societies are vital partners for pharmaceutical companies, shaping clinical practice through guidelines, education, and research. These societies, often composed of experts in various fields, contribute to the development of therapeutic standards, ensuring that new treatments meet the highest levels of clinical efficacy and safety.

Historically, medical societies have provided a platform for knowledge exchange, helping to bridge the gap between research and practice. Through conferences, congresses, and consensus papers, they play an essential role in disseminating scientific knowledge. For pharmaceutical companies, collaboration with medical societies offers a pathway to gain insight into the needs of the healthcare community, support clinical education, and contribute to the development of evidence-based guidelines.

The relationship between pharma companies and medical societies is collaborative but requires careful management to ensure that interactions are transparent and aligned with ethical standards. Medical societies, in turn, offer pharmaceutical companies the opportunity to engage with a broad spectrum of healthcare professionals, fostering scientific dialogue and advancing public health.

Identification and Development of External Experts

Criteria for Identifying External Experts

Pharmaceutical companies must apply rigorous criteria when identifying external experts. These experts must possess deep expertise in relevant therapeutic areas, demonstrated through a robust track record of academic qualifications, peer-reviewed publications, and research contributions. Leadership in clinical trials and participation in investigator-initiated studies are also key factors in the selection process, reflecting their ability to influence scientific and clinical practice.

Beyond academic credentials, external experts must demonstrate their impact within medical societies and regulatory bodies. Their influence can be seen through their role in shaping clinical guidelines, conducting high-impact research, and advocating for patient care. The integration of external experts' voices ensures that drug development is driven by real-world clinical needs and is reflective of the broader healthcare landscape.

Mapping and Segmentation of External Experts

Mapping and segmenting external experts is a critical step in maximizing their engagement. Experts are often categorized by their level of influence (national, regional, or global), allowing pharma companies to target individuals based on therapeutic needs, geography, and audience reach.

Advanced tools, such as customer relationship management (CRM) systems and AI-powered analytics, are instrumental in this process. These tools help companies analyze an expert's academic contributions, conference participation, and professional networks. By leveraging these technologies, companies can prioritize experts who align with their strategic goals and therapeutic priorities.

Segmentation is further refined by therapeutic areas and geographies. For example, companies may engage different experts depending on whether the focus is oncology, cardiology, or neurology. Similarly, geographic segmentation ensures that local market needs are addressed, fostering more targeted and meaningful interactions.

Engaging and Developing Relationships with External Experts

Engaging external experts begins with thoughtful, professional outreach. Initial contact should be tailored to the expert's interests and achievements, emphasizing shared goals and scientific collaboration. Relationship-building requires sustained efforts, focusing on mutual trust, transparency, and respect for the expert's expertise and time.

Long-term engagement strategies are crucial for maintaining productive collaborations. Regular interactions through advisory boards, clinical trial oversight, or scientific discussions foster deeper relationships. The involvement of Medical Science Liaisons (MSLs) and Field Medical Teams is particularly effective in managing these interactions, as they serve as a bridge between the pharmaceutical company and the external expert community.

Maintaining trust is paramount. Experts must feel that their input is valued and respected. Open communication, ethical conduct, and alignment of goals help foster a sustainable partnership.

Ethical Considerations and Transparency

Transparency and ethical conduct are central to successful engagement with external experts. Industry standards such as the EFPIA and PhRMA codes require companies to disclose financial relationships and ensure transparency in all interactions.

This includes clear contracts, disclosure of compensation, and regular audits to ensure compliance with regulatory guidelines.

Conflict of interest management is equally important. Companies must address any potential biases or conflicts in collaboration with external experts, ensuring that scientific integrity is maintained. External experts must be free to express their independent opinions, and pharma companies should avoid any undue influence that could undermine their credibility.

Roles of External Experts in Pharmaceutical Development

Contribution to Medicine Discovery and Preclinical Research

External experts contribute significantly to early-stage drug discovery. Their expertise in identifying therapeutic targets and validating biomarkers is invaluable for shaping the direction of preclinical research. These experts often serve on advisory boards that provide critical insights into scientific trends and emerging technologies, helping companies align their research with the needs of patients and clinicians.

Participation in Clinical Trials

External experts play a pivotal role in clinical trials, often serving as principal investigators or sub-investigators. Their insights are critical for trial design, particularly in selecting appropriate endpoints, patient populations, and safety parameters. Experts involved in Data Monitoring Committees (DMCs) provide ongoing oversight, ensuring patient safety and the integrity of the data collected.

Their involvement in clinical trials also enhances the credibility of the study. As recognized leaders in their field, their participation reassures stakeholders that the trial is scientifically rigorous and clinically relevant.

Advisory Boards

Advisory boards are another key area where external experts provide valuable input. These non-promotional, scientifically driven boards are designed to gather expert opinions on specific topics, such as protocol design, regulatory strategies, or market needs.

The agenda of advisory boards is focused and precise, with clear objectives and outcomes. Selection of board members is based on their expertise in the relevant therapeutic area, ensuring that discussions remain highly relevant and actionable.

Scientific Communication and Publications

External experts often collaborate on scientific communications, serving as co-authors of peer-reviewed articles, white papers, and abstracts. Their participation in medical congresses and symposia, where they present data or participate in expert panels, further disseminates key findings from clinical trials or research initiatives.

Media presence and contributions to medical education initiatives, such as Continuing Medical Education (CME), are other important avenues for expert involvement. These activities ensure that the scientific community stays informed of the latest advances in drug development and therapeutic innovations.

Collaboration with Key Medical Societies

Medical societies, such as the American Medical Association (AMA) and the European Society of Cardiology (ESC), play a fundamental role in shaping clinical guidelines and therapeutic standards. Their influence extends beyond the publication of guidelines to include educational programs, consensus papers, and ongoing research collaborations.

Pharma companies collaborate with medical societies to gain insights into clinical needs, ensure alignment with guideline developments, and engage in scientific discourse. This interaction fosters a robust exchange of ideas, driving evidence-based practice.

Strategic Collaboration with Medical Societies

Strategic partnerships with medical societies offer numerous benefits, including access to expert networks, clinical trial collaborations, and opportunities for educational outreach. Companies may sponsor society-led initiatives, such as clinical guideline updates or consensus conferences, providing financial support while ensuring non-promotional and unbiased outcomes.

Partnership models include sponsorship of research grants, advisory roles, and participation in educational symposia. These collaborations must be conducted transparently, with clear roles and responsibilities defined to maintain scientific independence and credibility.

Participation in Medical Society Events

Pharmaceutical companies frequently participate in medical society events, sponsoring major congresses, symposiums, and workshops. These events provide an opportunity to showcase clinical trial data, present scientific innovations, and foster networking with healthcare professionals and experts.

Organizing satellite symposia or expert panels during key events offers companies a platform to engage directly with medical professionals, facilitating scientific discussions in a controlled and ethical environment. Medical societies, in turn, play a crucial role in validating these engagements and ensuring they align with educational goals.

Expert Engagement and Management Strategies

Defining Roles and Responsibilities

Effective management of external experts begins with a clear definition of roles and responsibilities. External experts may serve on advisory boards, engage in educational roles such as webinars or conferences, or advocate for patient care and health policy. These roles must be carefully defined to ensure transparency and alignment with both the pharmaceutical company's objectives and the experts' expectations.

Advisory board participation is one of the most common forms of expert engagement. The selection of participants is based on their expertise in the therapeutic area and their ability to contribute to non-promotional, scientifically focused discussions. The setup and management of these boards require clear agendas, objectives, and expected outcomes. Each participant must understand the purpose of the board and the non-commercial nature of the discussions to ensure compliance with industry regulations.

In educational roles, external experts often serve as speakers at scientific events or lead webinars and workshops. These activities focus on providing healthcare professionals with unbiased, evidence-based information on therapeutic advancements. Experts also play a key role in facilitating professional development programs, helping to educate physicians and other stakeholders about new therapies and clinical practices.

Communication Channels and Tools

In the modern pharmaceutical landscape, digital engagement has become a crucial aspect of managing external experts. Virtual advisory boards, webinars, and telemedicine platforms are now commonly used to maintain ongoing interactions with experts across the globe. These digital platforms offer a more flexible and cost-effective means of communication, allowing for real-time collaboration without geographical limitations.

CRM systems and KOL databases are essential tools for tracking expert interactions. These systems allow pharmaceutical companies to monitor the contributions of external experts, assess the impact of their involvement, and manage future engagements. By analyzing data from these platforms, companies can identify which experts are most engaged, contribute the most value, and have the greatest influence in their respective fields.

Hybrid models that combine both in-person and virtual interactions are particularly relevant in a post-pandemic world. These models offer the flexibility of digital engagement while maintaining the personal touch of face-to-face meetings. For example, virtual advisory board meetings can be supplemented with in-person workshops or symposia at key medical events.

Metrics for Evaluating Expert Contributions

Evaluating the impact of external expert contributions is essential for ensuring the success of these collaborations. Pharmaceutical companies use both qualitative and quantitative performance indicators to assess the value that external experts bring to the table.

Quantitative metrics may include the number of scientific publications co-authored by experts, the number of clinical trials they are involved in, and the quality of the data they help generate. Trial outcomes, publication impact factors, and citation rates are useful measures of an expert's influence in their field.

Qualitative measures focus on the expert's adherence to clinical guidelines, their leadership in shaping therapeutic strategies, and their ability to engage in productive scientific discourse. These qualitative assessments can be tracked through CRM tools and engagement platforms, offering insights into the expert's level of influence and their contribution to shaping clinical practice.

Case Study: The Need for Enhanced Scientific Communication and Education on Biosimilars—A Call to Action for Medical Affairs (based on publication A. Krendyukov et al., "Current understanding, knowledge, and perception of biosimilars in a changing landscape of regulatory requirements", GaBI Journal)
Introduction

Biosimilars represent a significant advancement in the biopharmaceutical landscape, offering a cost-effective alternative to originator biologicals without compromising quality, safety, or efficacy. Despite their potential to reduce healthcare costs and expand patient access to life-saving therapies, the adoption of biosimilars remains suboptimal. This case study highlights the critical role of scientific communication and education, particularly from the Medical Affairs function in pharmaceutical companies, in fostering acceptance and trust among healthcare providers

(HCPs). Addressing knowledge gaps and misconceptions through evidence-based educational initiatives is essential to maximize the benefits of biosimilars for healthcare systems and patients alike.

Biosimilars: A Transformational Opportunity

Biological therapies have transformed the management of chronic and life-threatening conditions, including cancer, autoimmune diseases, and metabolic disorders. However, their high cost imposes a significant financial burden on healthcare systems, often limiting patient access. Biosimilars, developed to be highly similar to reference biologics, offer a solution to this challenge by providing equivalent therapeutic outcomes at a reduced cost.

A recent report by the Association for Accessible Medicine indicated that biosimilars are, on average, priced 50% lower than their reference products. Additionally, competition introduced by biosimilars has resulted in an average 25% price reduction for reference biologics. Despite these economic advantages and the rigorous regulatory requirements governing their approval, many HCPs remain hesitant to prescribe biosimilars due to misconceptions about their safety, efficacy, and interchangeability. Addressing these concerns is paramount to realizing the full potential of biosimilars.

Knowledge Gaps and Barriers to Acceptance

One of the primary barriers to biosimilar adoption is a lack of understanding among HCPs regarding the scientific principles underlying biosimilar development and approval. Unlike traditional generics, biosimilars are not identical copies of their reference products due to the inherent complexity of biological manufacturing. Instead, biosimilarity is established through a "totality of evidence" approach that includes extensive analytical characterization, comparative pharmacokinetic (PK) and pharmacodynamic (PD) studies, and limited clinical trials.

However, many HCPs erroneously prioritize clinical trial data over analytical and functional similarity assessments. In a survey of physicians across various specialties, 85% ranked clinical data as the most important evidence for biosimilar approval, while only 40% considered physicochemical data equally critical. This misunderstanding stems from a traditional reliance on randomized controlled trials (RCTs) as the gold standard for therapeutic evaluation, a paradigm that does not fully align with the scientific framework for biosimilars.

Moreover, confusion surrounding regulatory terms such as "extrapolation," "switching," and "interchangeability" further impedes acceptance. For instance, the concept of extrapolation—approving a biosimilar for multiple indications based on evidence generated in one indication—is often misunderstood as a compromise on safety or efficacy. Similarly, divergent regulatory approaches to interchangeability in regions such as the EU and the United States create uncertainty, undermining confidence in biosimilar use.

The Role of Medical Affairs in Driving Change

Medical Affairs teams within pharmaceutical companies are uniquely positioned to address these barriers through targeted educational initiatives and stakeholder engagement. As trusted partners to the medical community, Medical Affairs professionals can bridge the gap between scientific evidence and clinical practice, fostering an informed and confident approach to biosimilar utilization.

Key responsibilities of Medical Affairs in this context include the following:

- *Developing Evidence-Based Educational Materials*: Creating resources that clearly explain the "totality of evidence" framework, including the role of analytical comparability studies, PK/PD data, and immunogenicity assessments. Case studies and real-world data on successful biosimilar adoption can further illustrate their safety and efficacy.
- *Engaging with Key External Experts (KEEs)*: Collaborating with respected KEEs to disseminate accurate information about biosimilars through conferences, webinars, and peer-reviewed publications. KOLs can serve as advocates, reinforcing the credibility of biosimilars among their peers.
- *Conducting Training Programs for HCPs*: Organizing workshops and seminars tailored to specific therapeutic areas, addressing common misconceptions, and providing practical guidance on biosimilar prescribing and switching practices.
- *Facilitating Stakeholder Collaboration*: Encouraging dialogue among regulators, payers, and HCPs to harmonize terminology and streamline decision-making processes. This includes clarifying regional differences in regulatory standards for interchangeability and substitution.
- *Leveraging Digital Platforms*: Utilizing digital tools to reach a broader audience, including e-learning modules, interactive infographics, and mobile applications that provide up-to-date information on biosimilar developments.

Evolving Regulatory and Clinical Evidence

Regulatory agencies worldwide have continuously refined their guidelines to reflect advances in biosimilar science. For example, the World Health Organization's 2022 update to its biosimilar guidelines emphasizes the sufficiency of state-of-the-art analytical and PK/PD studies over traditional RCTs. Similarly, the UK Medicines and Healthcare products Regulatory Agency (MHRA) now recognizes that confirmatory efficacy trials are often unnecessary for biosimilar approval if robust analytical comparability is demonstrated.

Real-world evidence (RWE) also plays a pivotal role in building confidence. Studies have consistently shown no clinically meaningful differences in safety or efficacy between biosimilars and their reference products, even in sensitive patient populations. Publishing long-term pharmacovigilance data and post-marketing surveillance results can further reassure HCPs and patients about biosimilar reliability.

Overcoming Misconceptions About Switching and Interchangeability

The practice of switching between reference biologics and biosimilars—or among biosimilars—has been a contentious issue due to concerns about immunogenicity and loss of efficacy. However, multiple switching studies have demonstrated that such transitions do not compromise safety or therapeutic outcomes. For instance, an analysis of EU regulatory submissions revealed no adverse effects associated with single or multiple switches, validating the robustness of the biosimilar development process.

Harmonizing definitions and approaches to interchangeability across regions is essential to mitigate confusion. In the EU, interchangeability is inherent to biosimilar approval, whereas in the United States, additional clinical switch studies are required for a biosimilar to be designated as "interchangeable." Aligning these regulatory frameworks would simplify communication and promote global acceptance.

The Importance of Continuous Education

Biosimilar education should not be a one-time effort but a continuous process that evolves with scientific advancements and regulatory changes. Regular updates on emerging data, revised guidelines, and new therapeutic applications are essential to maintain HCPs' confidence and enthusiasm for biosimilars.

Educational efforts should also address the nuances of specific therapeutic areas. For example, oncology and immunology have been at the forefront of biosimilar adoption, offering valuable lessons for other specialties such as neurology, cardiology, and ophthalmology. Sharing these experiences can accelerate acceptance and prepare HCPs for the growing diversity of biosimilars entering the market.

Summary on Case Study

The successful integration of biosimilars into clinical practice hinges on the ability of Medical Affairs teams to provide clear, evidence-based, and ongoing education to HCPs. By addressing misconceptions, clarifying regulatory processes, and highlighting the rigorous scientific foundation of biosimilars, Medical Affairs can drive greater acceptance and utilization of these transformative therapies.

As patents for high-cost biologics continue to expire, the expansion of biosimilars offers a unique opportunity to enhance healthcare sustainability and improve patient access to innovative treatments. It is incumbent upon all stakeholders—pharmaceutical companies, regulators, payers, and HCPs—to collaborate in fostering a well-informed medical community that embraces the potential of biosimilars. Through strategic communication and education initiatives, we can ensure that biosimilars fulfill their promise of delivering high-quality, cost-effective care to patients worldwide.

Regulatory and Compliance Considerations

Regulatory Frameworks for Expert Engagement

Global regulatory frameworks govern the engagement of external experts in pharmaceutical companies. These frameworks ensure that interactions are transparent, ethical, and compliant with industry standards. In the United States, the Sunshine Act requires the disclosure of payments made to healthcare providers, including external experts, to prevent conflicts of interest. Similarly, in Europe, the EFPIA Disclosure Code outlines transparency requirements for interactions between pharmaceutical companies and healthcare professionals.

Compliance with US Food and Drug Administration (FDA), European Medicines Agency (EMA), and International Council for Harmonisation of Technical Requirements for Pharmaceuticals for Human Use (ICH) guidelines is critical in ensuring that expert collaborations do not influence the objectivity of clinical trials, drug approvals, or market access decisions. These guidelines outline the ethical boundaries of expert participation in advisory boards, clinical trials, and scientific publications, ensuring that collaborations remain focused on scientific integrity.

Conflict of Interest and Ethical Interactions

Pharmaceutical companies must manage financial relationships with external experts to avoid undue influence on their clinical or scientific opinions. Disclosure of honoraria, travel expenses, and other financial incentives is essential in maintaining transparency and upholding ethical standards.

To ensure that expert opinions remain independent, pharmaceutical companies must establish clear guidelines for conflict of interest management. These guidelines should be communicated to external experts from the outset, ensuring that their contributions are based on scientific evidence rather than financial incentives.

Pharma companies should also avoid creating situations where external experts may feel pressured to endorse a particular product or therapeutic strategy. Maintaining an open and transparent dialogue with experts, while respecting their autonomy, is key to fostering ethical interactions.

Legal Implications of Collaborating with Medical Experts and Societies

Collaborations with medical societies involve complex legal considerations, particularly around contracts, honoraria, and compensation models for experts. Contracts must clearly define the roles and responsibilities of both the pharmaceutical company and the external expert, outlining the scope of the collaboration and ensuring that both parties comply with regulatory and legal requirements.

Companies must also navigate the legal implications of cross-border collaborations, as regulations differ by country. Risk management strategies should be in place to ensure that all collaborations meet the legal standards of each region while maintaining the integrity of the scientific discourse.

Challenges and Solutions in Managing External Experts and Collaborations

Common Challenges

Pharmaceutical companies often face challenges when managing relationships with external experts. One common issue is the diversity of opinions among experts, which can lead to disagreements on clinical trial design, therapeutic strategies, or drug development priorities. Managing these diverse perspectives requires strong leadership and clear communication.

Regulatory hurdles can also pose significant challenges. Public perception of the pharmaceutical industry, particularly in the context of expert collaborations, can lead to scrutiny and skepticism. Ensuring transparency in financial relationships and ethical conduct is essential for maintaining trust with both the medical community and the public.

Additionally, cultural differences can complicate global expert management. Navigating these differences requires sensitivity to local practices, values, and regulatory landscapes.

Strategies for Overcoming Challenges

To address these challenges, pharmaceutical companies should prioritize continuous education and training for both internal teams and external experts. Ensuring that all parties are up-to-date on compliance and ethical standards is essential for mitigating risks and avoiding potential conflicts of interest.

Clear communication and conflict resolution strategies should be in place to manage differences of opinion among experts. Regular, transparent discussions can help foster mutual understanding and ensure that all stakeholders are aligned with the company's objectives.

Leveraging digital tools for global coordination can also help overcome logistical challenges, enabling real-time interactions with experts around the world. Virtual platforms offer an efficient way to manage relationships across different regions, ensuring consistent engagement and communication.

Clinical Cases and Real-World Examples

Collaborations with External Experts

Clinical cases from practice can illustrate how effective collaboration with external experts has led to breakthroughs in drug development and clinical practice. One notable example is the collaboration between a major pharmaceutical company and a group of oncology experts in the development of a novel immunotherapy. The external experts played a key role in shaping the trial design, selecting appropriate biomarkers, and ensuring that the therapy met regulatory requirements.

Another example is the development of clinical guidelines in cardiology, where external experts worked closely with both the pharmaceutical company and a major medical society to develop evidence-based recommendations for treating heart failure. This collaboration resulted in updated guidelines that improved patient outcomes and ensured that the new therapy was rapidly adopted into clinical practice.

Lessons Learned from External Expert and Society Interactions

Several lessons can be drawn from past collaborations with external experts and medical societies. One important takeaway is the need for early involvement of external experts in the drug development process. Early engagement allows experts to provide meaningful input on trial design, ensuring that studies are both scientifically rigorous and aligned with clinical needs.

Another key lesson is the importance of transparency. Clear communication around financial relationships and the non-promotional nature of advisory board discussions is critical for maintaining trust and credibility. Ensuring that all interactions are aligned with ethical guidelines prevents conflicts of interest and fosters a positive perception of the collaboration.

Future Trends in Expert Engagement and Collaboration

Emerging Trends in Expert Development

As digital tools continue to evolve, the role of AI-driven platforms in identifying and engaging experts will become increasingly important. These platforms use data analytics to map the influence of KEEs and predict emerging experts in niche therapeutic areas.

The globalization of medical expertise is another important trend. As pharmaceutical companies expand their operations into emerging markets, they must engage with a more diverse group of experts. This requires an understanding of local regulatory environments and healthcare needs, as well as the ability to foster relationships with experts from different cultural backgrounds.

The Future of Medical Societies in Pharma Collaborations

Medical societies will continue to play a pivotal role in shaping therapeutic strategies. The integration of real-world data (RWD) and RWE into expert collaborations will become increasingly important, providing more comprehensive insights into how therapies perform in clinical practice.

Incorporating patient-centric approaches is also gaining traction. Patient advocates and patient experts are becoming more involved in shaping medical strategies, offering unique perspectives on therapeutic development and patient care.

New Technologies and Digital Tools

The adoption of social media and virtual communities is transforming how external experts engage with the scientific community. Platforms like X (previously Twitter), LinkedIn, and ResearchGate are becoming important venues for scientific discourse, enabling experts to connect, share research, and influence medical practice in real time.

Telemedicine and virtual collaboration are also playing an increasingly significant role in expert engagement. Virtual advisory boards, online symposia, and telemedicine platforms enable ongoing communication with external experts, allowing pharmaceutical companies to maintain relationships without the need for frequent in-person meetings.

External experts are indispensable to the pharmaceutical industry, contributing to every stage of drug development, from preclinical research to post-marketing surveillance. Successful engagement with these experts requires clear communication, ethical conduct, and a commitment to transparency. Collaboration with medical societies further enhances the credibility of pharmaceutical innovations and ensures alignment with clinical guidelines and real-world needs.

Looking ahead, the future of expert engagement will be shaped by digital tools, AI-driven platforms, and an increasing focus on patient-centric approaches. As the pharmaceutical industry evolves, fostering meaningful and transparent relationships with external experts will remain a cornerstone of successful drug development.

Reference to Leading Industry Guidelines

Pharmaceutical companies operate in a highly regulated environment, particularly when it comes to engaging with external experts such as KOLs, KEEs, thought leaders, and SMEs. These collaborations are essential for advancing drug development, ensuring clinical trial integrity, and providing scientific communication to healthcare professionals and regulatory bodies. However, the complexity of these engagements requires adherence to strict regulatory frameworks to maintain transparency, integrity, and ethical conduct.

In this section, we explore the critical regulatory guidelines from global bodies such as the FDA, EMA, PhRMA, EFPIA, and other regional entities. We also present best practices from leading pharmaceutical companies to provide a practical perspective on compliance and transparency in expert collaboration.

Regulatory Guidelines Governing External Expert Collaboration

FDA Guidelines

The FDA sets the gold standard for regulatory oversight in the pharmaceutical industry, particularly in the United States. The FDA's focus on transparency, ethical conduct, and scientific integrity governs how companies engage external experts throughout the drug development lifecycle.

One key regulation from the FDA is the Sunshine Act, part of the Physician Payments Sunshine Act under the Affordable Care Act (ACA). This legislation mandates the disclosure of financial relationships between pharmaceutical companies and HCPs, including external experts. Pharmaceutical companies are required to report any payments, gifts, or services provided to external experts, ensuring full transparency to the public.

- *Link to the Physician Payments Sunshine Act*: https://www.cms.gov/OpenPayments

The Sunshine Act's primary goal is to prevent conflicts of interest that could arise from financial compensation, ensuring that external expert contributions remain objective and science-driven. Pharmaceutical companies must ensure that their payments to experts do not influence clinical trial results, guideline development, or scientific publications.

Best Practices for FDA Compliance

- *Regular Audits and Disclosures*: Leading pharmaceutical companies conduct regular internal audits to ensure compliance with the Sunshine Act. They maintain detailed records of all interactions with external experts, which are disclosed publicly.
- *Advisory Board Compliance*: Companies ensure that advisory board discussions are non-promotional and focused solely on scientific content. By keeping these boards compliant with FDA regulations, companies avoid any perception of marketing influence.

For example, *Company Y* has instituted robust compliance measures, regularly auditing payments to external experts and providing real-time access to its disclosure records. They also ensure advisory board members are compensated fairly, transparently, and in accordance with Sunshine Act guidelines.

EMA Guidelines

The EMA operates similarly to the FDA but within the European Union's jurisdiction. It governs expert collaborations through strict guidelines that emphasize transparency, impartiality, and scientific rigor.

A notable EMA regulation is the EFPIA Disclosure Code, which mirrors the Sunshine Act in terms of transparency. The EFPIA Disclosure Code requires pharmaceutical companies to disclose payments to healthcare professionals and organizations, including external experts, on a public platform.

- *Link to EFPIA Disclosure Code*: https://www.efpia.eu/

In addition to the EFPIA Disclosure Code, the ICH E6 Good Clinical Practice (GCP) Guidelines are critical in ensuring ethical conduct in clinical trials. These guidelines highlight the roles of external experts as investigators and scientific advisors, emphasizing their responsibility to ensure patient safety, trial integrity, and unbiased data interpretation.

Best Practices for EMA Compliance

- *Transparent Compensation Models*: Pharmaceutical companies operating in the EU adhere to compensation models that reflect fair-market value for expert contributions. Compensation is strictly for scientific advice, clinical trial oversight, or publication support, avoiding any form of promotional influence.

- *EFPIA Compliance Audits*: Leading companies conduct routine audits to ensure compliance with the EFPIA Disclosure Code. Their efforts in establishing robust systems for tracking interactions with external experts ensure that all payments are disclosed accurately.

For instance, *Company Q* adheres to a comprehensive process for handling payments to external experts, ensuring that all interactions are documented and publicly disclosed, as mandated by EFPIA. This transparency fosters trust among healthcare professionals, regulators, and the public.

PhRMA Guidelines

PhRMA represents leading biopharmaceutical research companies in the United States and provides voluntary guidelines through its *PhRMA Code on Interactions with Healthcare Professionals*. The PhRMA Code governs interactions between pharmaceutical companies and healthcare professionals, including external experts, to ensure that these relationships are ethical and non-promotional.

- *Link to PhRMA Code*: https://www.phrma.org

The PhRMA Code emphasizes that all collaborations with external experts must be science-driven, ensuring that any compensation provided is reasonable and reflective of the expert's professional time and effort. Companies must also avoid any activities that could be perceived as influencing the expert's professional judgment or clinical practice.

Best Practices for PhRMA Compliance

- *Non-Promotional Engagement*: To remain compliant with PhRMA guidelines, companies ensure that advisory board meetings, speaker engagements, and clinical trial oversight by external experts are entirely non-promotional. This separation of scientific and promotional activities is crucial in maintaining the integrity of expert contributions.
- *Clear Contractual Agreements*: Leading pharmaceutical companies ensure that every engagement with an external expert is governed by a clear, transparent contract. These contracts outline the nature of the collaboration, expected deliverables, and compensation, ensuring compliance with PhRMA standards.

ICH Guidelines

The ICH guidelines, specifically ICH E6 (R2/R3) on Good Clinical Practice (GCP), are among the most widely adopted regulatory frameworks for clinical trials globally. These guidelines emphasize that expert involvement in clinical trials must be conducted with the highest standards of ethics and patient safety in mind.

- *Link to ICH Guidelines*: https://www.ich.org/page/guidelines

The ICH E6 guidelines outline the responsibilities of investigators, including external experts, who act as principal investigators or advisors on clinical trials. Their role in trial design, patient safety oversight, and data interpretation is critical to ensuring that the trial adheres to the highest ethical and scientific standards.

Best Practices for ICH Compliance

- *Ethical Trial Design*: Leading pharmaceutical companies can ensure that external experts involved in clinical trial design are well-versed in ICH GCP guidelines. Their contributions to clinical protocol design focus on patient safety and data integrity, ensuring that trials meet both regulatory and ethical standards.
- *Data Integrity and Oversight*: Pharmaceutical companies can exemplify best practices by engaging external experts to serve on Data Monitoring Committees (DMCs), ensuring that clinical trials are reviewed for patient safety and efficacy throughout the study. These experts are independent and bound by ICH standards for data oversight.

Case Study: ABPI's Role and ABPI Code Evolution

The Association of the British Pharmaceutical Industry (ABPI) represents research-based biopharmaceutical companies in the United Kingdom, which collectively drive innovation in medicine, vaccines, and biotechnology. Acting as a bridge between pharmaceutical companies, regulators, HCPs, and the broader healthcare system, ABPI plays a pivotal role in fostering collaboration, advancing public health initiatives, and ensuring ethical practices across the industry.

ABPI's influence extends beyond advocacy; it also establishes guidelines for the ethical promotion of pharmaceutical products through its Code of Practice. This ensures compliance with UK laws and alignment with global best practices, safeguarding public trust. The ABPI Code of Practice is administered by the Prescription Medicines Code of Practice Authority (PMCPA) and is pivotal in setting high standards for marketing, interactions with HCPs, and patient engagement.

Key Aspects of the ABPI Code of Practice and Recent Updates

The 2024 updates to the ABPI Code of Practice, effective October 1, 2024, reflect a commitment to maintaining ethical rigor and adapting to evolving industry needs.

The most significant revisions pertain to prescribing information (PI), interactions with HCPs, and the integration of emerging technologies. Below is a summary of key changes:

1. *Prescribing Information (Clause 12)*
   The updates to Clause *12* streamline the presentation of PI while ensuring accessibility and compliance:

- *Digital Enhancements*: PI can now be delivered via QR codes, provided they are clear, prominent, and direct users to the most up-to-date information. Companies are encouraged to use dynamic QR codes to maintain accuracy.
- *Accessibility Requirements*: Promotional materials must include a clear statement about PI availability, visible at first glance without requiring scrolling or page-turning.
- *Multi-Product PI Clarity*: When advertising multiple medicines, PI must be easily distinguishable for each product, avoiding scenarios where users need to navigate through irrelevant details.

2. *Governance and Documentation*
   The Code emphasizes enhanced governance in interactions with HCPs and other stakeholders:

- *Support Documentation (Clause 10.4)*: New requirements mandate a rationale and written agreements for supporting HCPs to attend events. This change stems from audit findings highlighting governance risks.
- *Disclosure Transparency (Clause 28.1)*: Companies must disclose financial interactions with HCPs, organizations, and patient groups via standardized templates on the Disclosure UK platform.
- *Collaborative Working (Clause 20.3)*: Updated guidelines highlight the importance of future-proofed texts and alignment with ABPI collaboration standards.

3. *Social Media and Digital Advertising*
   Recognizing the evolving digital landscape, the 2024 Code incorporates:

- *Social Media Policies*: Companies must develop policies for both personal and business use of social media by employees.
- *Clarity in Digital Reprints*: Digital materials must ensure PI accessibility through a direct, single-click link, maintaining consistency with print standards.

4. *Ethical Standards and Market Research (Clause 5)*

   - Companies are now required to explicitly communicate their high ethical standards through policies and training. Market research materials must state they are "commissioned," not "sponsored," by pharmaceutical companies to avoid ambiguity.

Table Comparison with EFPIA and PhRMA Codes of Conduct

To place the ABPI Code in a broader context, a comparison with the EFPIA Code and the PhRMA Code reveals similarities and distinct approaches to ethical governance and compliance.

Aspect	ABPI	EFPIA	PhRMA
Interactions with HCPs	Transparency in financial interactions with clear disclosures via Disclosure UK; written agreements required for event support.	Similar transparency with national-level disclosures; emphasizes cross-border transparency.	Limits gifts and hospitality, focuses on educational support, and prohibits undue influence on HCPs.
Patient Engagement	Collaboration with patient organizations under strict compliance, ensuring transparency.	Requires disclosure of financial and in-kind contributions to patient organizations.	Emphasizes clear communication of financial relationships and patient-centric advocacy.
Digital and Technological	Integrates digital solutions such as QR codes for PI, ensuring accessibility and compliance.	Gradually integrating digital solutions, though less comprehensive than ABPI.	Slow adoption of digital technologies; guidelines less specific compared to ABPI's detailed approach.
Governance	Strong emphasis on written documentation, governance, and disclosure of financial contributions.	Aligns with EU-wide governance policies, focusing on transparency and harmonization.	Governance primarily focuses on the US regulatory environment, emphasizing compliance with FDA regulations.
Market Research	Requires explicit communication that research is "commissioned" by pharmaceutical companies to avoid ambiguity.	Similar standards for transparency in sponsorship versus commissioning.	Strong emphasis on ethical considerations but less detailed than ABPI on commissioning vs. sponsorship clarity.

Conclusion and Future Outlook

The 2024 ABPI Code of Practice embodies a proactive approach to ethical compliance, balancing regulatory rigor with the dynamic needs of modern healthcare. Its emphasis on transparency, digital integration, and collaborative governance aligns with global trends while addressing unique national requirements.

Looking ahead, the pharmaceutical industry will likely witness further harmonization of ethical codes across regions, driven by globalization and the rise of digital health technologies. Enhanced collaboration between organizations like ABPI,

EFPIA, and PhRMA could establish a unified global framework, ensuring consistent ethical standards worldwide.

As the industry continues to evolve, the ABPI remains a beacon of innovation and ethical leadership, setting benchmarks that inspire trust and drive progress in healthcare.

Case Study: Best Practices from Leading Pharmaceutical Companies

Company X

Company X has developed a comprehensive framework for managing external expert relationships, ensuring adherence to EFPIA guidelines in Europe and other relevant regulatory frameworks worldwide. The company is known for its rigorous approach to ensuring transparency in financial interactions with external experts, particularly in clinical trials and advisory roles.

One of Company X's standout practices is its robust use of digital platforms to track KOL engagement, monitor their scientific contributions, and disclose interactions publicly. By using advanced CRM systems and ensuring compliance with regional regulations, Company X maintains trust and credibility within the medical community.

Company Y

Company Y is renowned for its transparent and strategic approach to external expert collaborations. The company regularly engages with KOLs through advisory boards, clinical trial oversight, and scientific publications. Company Y ensures compliance with the Sunshine Act and PhRMA Code, maintaining full disclosure of payments and ensuring that all expert contributions are evidence-based and non-promotional.

Company Y's best practice in expert engagement involves conducting regular audits of financial interactions with KOLs and adhering strictly to the fair market value compensation model. This approach prevents any potential conflict of interest, ensuring that experts' input remains scientifically sound and unbiased.

Company Z

Company Z has established a leading model for expert collaboration, guided by a strict adherence to *PhRMA* guidelines and global regulatory standards. The company emphasizes clear contractual agreements with external experts, ensuring that all collaborations are science-driven and focused on therapeutic advancements.

Company internal compliance teams ensure that all advisory board meetings and expert engagements are non-promotional, and compensation is provided in a transparent and regulated manner. By maintaining clear boundaries between commercial and scientific interactions, Company Z sets a high standard for ethical expert collaboration.

Company Q

Company Q places a strong emphasis on transparency and ethical conduct when engaging with external experts. The company's compliance with the EFPIA

Disclosure Code ensures that all payments to healthcare professionals, including external experts, are fully disclosed. Company Q uses a comprehensive tracking system to document every interaction with experts, ensuring that these collaborations are fully compliant with regulatory standards.

Company Q also emphasizes the importance of involving external experts early in the drug development process. By engaging KOLs during the preclinical and clinical trial stages, the company ensures that its research aligns with clinical needs and patient outcomes.

Summary

Collaborations between pharmaceutical companies and external experts are essential to advancing therapeutic innovation, but they must be conducted within a strict regulatory framework to ensure transparency, ethical conduct, and scientific integrity. Regulatory bodies such as the FDA, EMA, and PhRMA have established clear guidelines that govern these interactions, emphasizing transparency, conflict of interest management, and non-promotional activities.

Best practices from leading pharmaceutical companies such as *Company X*, *Company Y*, *Company Z*, and *Company Q* demonstrate the importance of maintaining rigorous compliance with these regulatory frameworks. These companies have developed robust systems for tracking interactions, ensuring transparency, and engaging external experts in a way that aligns with both regulatory requirements and ethical standards.

As the pharmaceutical industry continues to evolve, the role of external experts will remain critical to the success of drug development. By adhering to leading industry guidelines and best practices, pharmaceutical companies can ensure that their collaborations with external experts are both productive and compliant.

Engaging External Experts Throughout the Product Life Cycle

Advisory boards are one of the most effective and common ways pharmaceutical companies interact with KEEs. These boards are strategic, non-promotional, and scientifically focused platforms designed to gather expert insights, address critical questions, and guide decision-making at various stages of a product's life cycle. From early-stage drug development to post-marketing, advisory boards offer pharma companies a unique opportunity to engage with experts, ensuring that their therapeutic strategies align with clinical realities and unmet needs.

This case study explores the key objectives of advisory boards at different stages of the product life cycle, the organizational aspects essential for success, common pitfalls to avoid, and best industrial practices. We will also delve into the integration of technology in modern advisory board management, including the rise of virtual ad boards, automation tools for documentation, and digital platforms for enhancing expert collaboration.

Key Objectives of Advisory Boards Across the Product Life Cycle

Advisory boards play a crucial role at different stages of a product's life cycle, offering pharmaceutical companies the opportunity to gain specific, actionable insights that drive both the development and commercialization of new therapies. Below is an exploration of how advisory board objectives vary across the product's journey, from early development to post-launch.

Preclinical and Early Development Stages

At the early stages of product development, the primary objective of advisory boards is to gather expert input on therapeutic targets and validate early-stage research. External experts, often with deep clinical or academic expertise, provide valuable insights into the scientific feasibility of potential drug candidates and biomarkers.

Key objectives in these early stages include the following:

- *Understanding the Therapeutic Landscape*: Experts help pharma companies understand the current treatment options, unmet needs, and patient populations that could benefit from a new therapeutic approach.
- *Evaluating Preclinical Data*: External experts provide feedback on preclinical data, helping refine the scientific direction and offering insights into potential challenges.
- *Protocol Design Input*: In some cases, especially in areas like oncology or rare diseases, experts are consulted to shape early-stage clinical protocols to ensure that future trials will generate meaningful data.

Clinical Development and Regulatory Submission

As products move into clinical development, advisory boards become critical for optimizing clinical trial design, addressing regulatory requirements, and ensuring that development programs are aligned with real-world clinical practice.

Key objectives at this stage include the following:

- *Refining Clinical Trial Design*: External experts provide guidance on patient selection, appropriate endpoints, and trial duration. Their insights help ensure

that the clinical trial is scientifically rigorous and aligned with the therapeutic area's needs.
- *Addressing Safety and Efficacy Questions*: Experts offer critical input on potential safety concerns and suggest ways to mitigate risks, as well as insights into efficacy parameters.
- *Preparing for Regulatory Submissions*: Advisory boards can help prepare for regulatory submissions by ensuring that clinical data packages meet the expectations of regulatory authorities like the FDA and EMA.

Late-Stage Development and Pre-Launch

As a product approaches regulatory approval, the role of advisory boards shifts toward strategic insights for commercialization. External experts guide pharma companies on how to position the product in a competitive therapeutic landscape.

Key objectives at this stage include the following:

- *Market Positioning and Competitive Landscape*: Experts offer insights into the current treatment landscape, identifying where the new product will fit and how it compares to existing therapies.
- *Health Economics and Outcomes Research (HEOR)*: Advisory boards can help develop strategies around RWE generation and health economic studies that will support pricing, reimbursement, and market access.
- *Guideline Development and KOL Endorsement*: At this stage, pharma companies work with experts to ensure that the new therapy is included in clinical guidelines and that KOLs in the field are aware of the product's value.

Post-Launch and Life Cycle Management

Post-launch, advisory boards remain critical for ensuring the long-term success of the product. They are used to gather feedback from clinical practice, monitor safety and efficacy in the real world, and explore opportunities for line extensions or new indications.

Key objectives in the post-launch phase include the following:

- *Monitoring RWE*: Experts provide feedback on the drug's performance in everyday clinical practice, helping identify any safety concerns or opportunities for improvement.
- *Evaluating New Indications*: Advisory boards can help assess potential new therapeutic indications or patient subgroups where the product could be effective.

- *Feedback on Market Access and Reimbursement*: Ongoing discussions focus on securing reimbursement and addressing payer concerns, particularly as real-world data becomes available.

Key Organizational Aspects of Advisory Boards

A well-organized advisory board ensures productive and efficient collaboration with external experts. Success hinges on careful planning, clear objectives, and precise execution. Here are key organizational elements to consider when conducting advisory boards:

Agenda Setting

The agenda must be *focused, clear, and specific* to the objectives of the meeting. Defining the key questions and topics to be discussed is crucial to avoid scope creep and to ensure that the experts remain focused on delivering actionable insights. Agendas should be prepared well in advance, distributed to the participants, and aligned with the company's strategic goals.

- *Pre-Meeting Materials*: Experts should receive pre-read materials that provide necessary background information, clinical data, or key questions. This allows them to come prepared with informed opinions.
- *Moderator's guide to facilitate and to guide the meeting and interactive conversation* by keeping discussions focused, ensuring adherence to the meeting objective and agenda, and allowing all experts to contribute.

Selection of Participants

Choosing the right experts is critical to the success of an advisory board. Participants should have relevant expertise in the therapeutic area and offer diverse perspectives (e.g., clinicians, researchers, payers, or patient advocates). The selection process should be transparent, and participants should be briefed on the specific goals of the advisory board.

- *Diversity in Expertise*: It is important to include experts with different specialties, geographies, and clinical perspectives to ensure a well-rounded discussion.

Non-Promotional and Objective Focus

Advisory boards must remain *non-promotional* and focused solely on scientific discussions. To comply with regulatory requirements such as the *Sunshine Act, the PhRMA Code*, and *the EFPIA Disclosure Code*, the objectives of the advisory board must be purely scientific or clinical, with no promotional intent.

- *Compliance and Documentation*: It is important to document that the advisory board is non-promotional and the compensation provided is for scientific expertise. All interactions should comply with regulatory requirements for transparency.

Logistics and Planning

Proper logistics planning ensures that the meeting runs smoothly. This includes arranging travel, accommodation, and meeting space (or virtual platforms). Attention should also be given to the timing of the meeting to accommodate experts' schedules.

- *Hybrid and Virtual Advisory Boards*: The shift toward hybrid or fully virtual advisory boards is increasing. Virtual boards provide flexibility, allowing experts from different geographies to participate without the need for travel. Companies must ensure that the technology platform used is secure, reliable, and conducive to collaboration.

Post-Meeting Follow-Up

The effectiveness of an advisory board does not end when the meeting concludes. Post-meeting activities, such as generating meeting minutes, summarizing key takeaways, and communicating outcomes to participants, are essential. Minutes should capture critical insights and recommendations, and they should be shared with internal stakeholders to inform decision-making.

- *Timely Feedback and Implementation*: Follow-up on the advisory board's recommendations is critical. Companies should track the implementation of expert feedback in product strategies or clinical trial designs.

Common Pitfalls to Avoid

While advisory boards are valuable, several potential pitfalls can undermine their effectiveness:

- *Lack of Focused Objectives*

Without clear and focused objectives, advisory boards can become unfocused and fail to generate actionable insights. Overloading the agenda with too many topics or allowing discussions to veer off track diminishes the value of expert contributions.

Solution: Set specific, measurable objectives before the meeting and ensure that the agenda is designed to address these objectives directly.

- *Overly Promotional Tone*

Experts may feel uncomfortable if the advisory board seems focused on product promotion rather than scientific discourse. This can damage the relationship between the pharma company and the experts, leading to a loss of trust.

Solution: Ensure that the advisory board remains purely scientific and non-promotional. Maintain a clear boundary between marketing activities and scientific discussions.

- *Inadequate Participant Engagement*

If participants are not properly engaged, the advisory board may fail to generate meaningful insights. This can occur when participants are not well-prepared or if the discussion is dominated by a few individuals.

Solution: Distribute pre-meeting materials and ensure that the moderator encourages participation from all attendees. Diverse perspectives should be actively solicited.

- *Poor Documentation and Follow-Up*

Failing to document key insights or not following up on expert recommendations can waste the valuable input provided during the advisory board. Without proper follow-up, the impact of the advisory board on decision-making may be diminished.

Solution: Assign a dedicated team to record meeting minutes, capture insights, and ensure that expert recommendations are incorporated into strategic planning.

Best Practices for Conducting Advisory Boards

Pharmaceutical companies that excel in organizing advisory boards share several best practices:

- *Clear and Concise Agenda*: The best advisory boards have a clearly defined agenda that focuses on the key objectives of the meeting. Leading companies distribute pre-meeting materials and ensure that all participants are aligned with the purpose of the discussion.

- *Engaging and Diverse Expert Participation*: Successful advisory boards include a well-balanced group of experts from diverse fields. Most of the companies are known for their comprehensive selection processes, which ensure that their advisory boards reflect a broad spectrum of perspectives.
- *Integration of Technology for Virtual Advisory Boards*: With the increasing adoption of digital tools, some companies have integrated advanced platforms for conducting virtual advisory boards. These platforms include capabilities for real-time collaboration, recording sessions, and generating automated minutes. Virtual advisory boards provide greater flexibility, enabling companies to engage with experts from different regions without the logistical constraints of in-person meetings.

Integration of Technology in Advisory Boards

Virtual Advisory Boards

In recent years, the use of virtual advisory boards has surged due to the global pandemic and the need for more flexible, cost-effective interactions. Virtual platforms allow companies to convene experts from around the world without the need for travel or physical presence. This increases participation and allows for more frequent engagements.

Advantages of Virtual Advisory Boards

- *Cost Efficiency*: Reduced travel and accommodation costs make virtual advisory boards more cost-effective.
- *Global Reach*: Virtual platforms allow for the inclusion of a broader range of experts from different geographies.

Automated Documentation Tools

Technology also plays a key role in streamlining the documentation of advisory board meetings. Automated tools can record sessions, generate minutes, and organize insights for future reference.

- *Meeting Minute Automation*: Software tools can automatically transcribe meetings, reducing the need for manual note-taking and ensuring that all critical points are captured accurately.

- *Centralized Data Repositories*: Companies are increasingly using centralized digital repositories to store advisory board documents, making it easy for internal teams to access expert insights and recommendations.

In summary, advisory boards remain one of the most strategic and effective ways for pharmaceutical companies to engage with external experts. By tailoring objectives to the different stages of a product's life cycle, organizing meetings with clear agendas, and leveraging technology for virtual collaboration, pharma companies can harness the expertise of KOLs to inform critical decisions. Avoiding common pitfalls, such as promotional bias or poor documentation, ensures that advisory boards deliver valuable, actionable insights. Looking forward, the integration of digital platforms and automation tools will continue to enhance the efficiency and effectiveness of advisory boards in the pharmaceutical industry.

Part III

Chapter 9
Medical Affairs Enabling Activities

Key Highlights
- Medical Affairs ensures alignment between scientific data and business objectives while upholding ethical standards.
- Provides scientific expertise for product launches, market access, and regulatory compliance.
- MSLs facilitate engagement with HCPs and contribute to RWE generation.
- Digital transformation and AI enhance data analytics and stakeholder engagement.
- Post-market surveillance and pharmacovigilance activities safeguard patient safety.

Medical Affairs has evolved into a central pillar in the pharmaceutical industry, spanning scientific, regulatory, and commercial functions. Its strategic role now encompasses more than just a support mechanism for clinical development—it has become an enabler of business, helping to navigate complex landscapes while ensuring the scientific and ethical integrity of decisions. This chapter explores the critical enabling activities Medical Affairs provides to support business development, commercial strategy, regulatory compliance, and sales, as well as its contributions to innovation, corporate reputation, and cross-functional collaboration.

Strategic Integration of Medical Affairs into Business Functions

Medical Affairs has emerged as a business enabler that fosters collaboration between commercial and scientific functions, ensuring that scientific data is integrated into key business decisions while *maintaining the highest standards of scientific integrity.*

Medical Affairs is uniquely positioned to bridge the gap between commercial functions, such as marketing and sales, and the scientific, regulatory, and ethical

A. Krendyukov, *Medical Affairs' Fundamentals: The Next Generation*,
https://doi.org/10.1007/978-3-031-92588-7_9

considerations that underpin pharmaceutical products. By providing expert insights, Medical Affairs helps ensure that business decisions align with clinical evidence, enhancing both product credibility and market success.

Supporting Commercial Strategy

- *Scientific Expertise for Product Launches*

  One of the key ways in which Medical Affairs supports commercial strategy is through its involvement in product launches. Medical Affairs teams provide the necessary clinical insights and data that ensure a product's readiness for market entry. This includes validating the efficacy and safety data from clinical trials, helping to shape the scientific narrative that will resonate with HCPs and regulators, and supporting the development of medical education programs.
- *Building Scientific Narratives*

  Medical Affairs plays a crucial role in brand strategy by constructing scientifically rigorous narratives that align with both clinical and market realities. By communicating the clinical benefits of a product clearly and credibly, Medical Affairs contributes to messaging that supports both regulatory approval and commercial success. This narrative must be built on robust evidence and tailored to different stakeholders, from regulators to HCPs and payers.
- *Business Development and Licensing Support*

  In the context of mergers, acquisitions, or partnerships, Medical Affairs performs due diligence by assessing the clinical viability of potential products or technologies. This evaluation involves scrutinizing clinical trial data, understanding the competitive landscape, and evaluating the potential for market success based on scientific evidence. This is a key part of business development, as it ensures that acquisitions or in-licensing opportunities are grounded in sound scientific data.

  Medical Affairs also plays a vital role in establishing and maintaining strategic partnerships with external entities, including academic institutions, research organizations, and clinical experts. These collaborations often provide critical insights and data that drive innovation and facilitate market access.

  Medical Affairs collaborates closely with Regulatory Affairs to ensure that the clinical data supporting a product is aligned with regulatory expectations. During the pre-approval phase, Medical Affairs contributes to dossier preparation by providing data that supports both the efficacy and safety of the product. Post-approval, Medical Affairs continues to play a vital role by monitoring the product's performance in real-world settings, contributing to regulatory filings, and responding to regulatory inquiries.

  Medical Affairs ensures that all scientific data included in regulatory submissions is accurate and compliant with global regulations. By collaborating with

Regulatory Affairs, Medical Affairs helps to develop product labeling that accurately reflects the clinical data, ensuring compliance with local and global guidelines.

- *Sales and Market Access Collaboration*
  Medical Science Liaisons (MSLs) are key members of Medical Affairs who engage directly with HCPs and provide scientific education on the company's products. MSLs serve as conduits between internal teams, sales, and external stakeholders, delivering scientific data to physicians and collecting real-world insights that feed back into Medical Affairs and business strategy.
- *Medical Affairs and Health Economics*
  Health economics plays an increasingly important role in the pharmaceutical industry, particularly in ensuring market access. Medical Affairs collaborates with health economics teams to support payer negotiations, providing scientific evidence that demonstrates the value and cost-effectiveness of a product. This data is critical for reimbursement strategies and gaining access to national formularies or health insurance plans.

Medical Affairs as a Key Driver for Innovation

Medical Affairs contributes significantly to innovation within the pharmaceutical industry by providing insights derived from RWE, supporting research and development, and leveraging digital tools and data analytics to drive forward clinical and business goals.

Post-Market Surveillance and Real-World Evidence (RWE)

Once a product is launched, Medical Affairs takes on a crucial role in post-market surveillance by collecting and analyzing RWE. This data is instrumental in shaping ongoing business decisions, driving label expansions, and supporting safety monitoring. By understanding how a drug performs outside the clinical trial setting, Medical Affairs can help refine treatment guidelines and product positioning in the market.

Supporting R&D Through Feedback Loops

Medical Affairs is a key conduit between clinical practice and Research and Development (R&D). Insights gathered from HCPs through Medical Affairs teams feed back into the R&D pipeline, helping to refine existing products, identify new areas for development, and improve clinical trial designs. This feedback loop helps the company stay agile and responsive to market needs.

Digital Transformation in Medical Affairs

Data Analytics and Artificial Intelligence
With the increasing availability of big data, Medical Affairs is leveraging data analytics and artificial intelligence (AI) to extract clinical insights from vast datasets. These tools enable more precise decision-making and allow for the identification of trends or patterns that may not be apparent through traditional analyses.

Digital Engagement and Virtual Medical Affairs
The use of digital platforms has transformed how Medical Affairs interacts with stakeholders. Virtual engagement tools allow for broader outreach to HCPs, providing education and scientific data through webinars, online platforms, and digital conferences. This not only expands the reach of Medical Affairs but also facilitates more rapid dissemination of critical scientific information.

Medical Affairs as the Guardian of Scientific Integrity

While Medical Affairs plays an important role in enabling business functions, it is also the primary guardian of scientific integrity within a pharmaceutical company. This balance is crucial for maintaining trust and ensuring that commercial objectives are grounded in sound scientific evidence.

Balancing Business Objectives with Ethical Standards

Medical Affairs ensures that business decisions are backed by solid scientific evidence and that these decisions are made within ethical boundaries. This is particularly important when balancing the pressures of commercial objectives with the need to prioritize patient safety and clinical efficacy.

Compliance with Global Regulations and Guidelines

As the pharmaceutical industry is highly regulated, Medical Affairs plays a critical role in ensuring compliance with a variety of global regulations, including those from the US Food and Drug Administration (FDA), European Medicines Agency (EMA), and International Council for Harmonisation-Good Clinical Practice (ICH-GCP). This ensures that clinical trials and product marketing are conducted ethically and that all data dissemination adheres to strict guidelines.

Transparency and Scientific Rigor in Communications

Scientific Publications and Data Dissemination

Medical Affairs is responsible for the transparent communication of scientific data through peer-reviewed publications, medical conferences, and other platforms. This dissemination must be handled with scientific rigor to ensure that the data is accurate, unbiased, and ethically presented.

Engagement with Healthcare Professionals

One of the core functions of Medical Affairs is engaging with HCPs to provide accurate and balanced scientific information. These engagements must be transparent and focused on education rather than promotion, helping to maintain the credibility of both the company and its products.

Medical Affairs: Enhancing Corporate Reputation and Trust

Beyond its technical and regulatory functions, Medical Affairs plays an important role in enhancing the company's corporate reputation by building trust with external stakeholders, including key opinion leaders (KOLs), patient advocacy groups, and the public.

Building Credibility with External Stakeholders

Key External Experts (KEEs)
Medical Affairs is often responsible for cultivating relationships with KEEs and facilitating advisory boards. These partnerships help shape product development and marketing strategies by incorporating expert opinions into the decision-making process.

Public and Patient Advocacy Groups
In addition to working with KEEs, Medical Affairs also engages with public and patient advocacy groups. This fosters transparency and ensures that patient needs and concerns are considered during the development and commercialization of new products.

Risk Mitigation and Crisis Management

Product Safety and Pharmacovigilance
Medical Affairs is closely involved in pharmacovigilance activities, ensuring that any safety concerns related to a product are promptly addressed. By managing safety data and responding to adverse events, Medical Affairs helps to mitigate risks and maintain product safety.

Handling Controversial Data or Safety Concerns
In the event of a product recall or safety issue, Medical Affairs plays a key role in managing the crisis. By communicating clearly and transparently with stakeholders, Medical Affairs helps to maintain trust in the company and its products during times of uncertainty.

Medical Affairs: Additional Enabling Activities

Medical Affairs plays a pivotal role in pharmaceutical companies, not just as a scientific advisor, but as an integral part of multiple enabling activities that support business operations and compliance. This section explores some of the key enabling activities of Medical Affairs, including participation in promotional and non-promotional review committees, handling medical information requests, and supporting cross-functional teams. Each of these activities, while distinct in function, collectively ensures that scientific integrity, compliance, and strategic alignment are maintained across business operations.

Promotional Review Committees

Definition Promotional Review Committees (PRCs) are formal bodies within pharmaceutical companies responsible for reviewing and approving promotional materials, ensuring that they are scientifically accurate, compliant with regulatory standards, and ethically sound. These materials can include advertisements, brochures, website content, and other marketing collateral directed at HCPs or consumers.

Purpose The primary purpose of the PRC is to safeguard against the dissemination of misleading or inaccurate information about pharmaceutical products. Regulatory agencies, such as the FDA and EMA, have strict guidelines governing promotional claims made by pharmaceutical companies. These guidelines aim to protect public health by ensuring that promotional content is based on solid scientific evidence and does not exaggerate the benefits or downplay the risks of a product.

Medical Affairs plays a crucial role in the PRC by providing the necessary scientific input to ensure that promotional materials are factually correct and consistent with the latest clinical data. Typically, the PRC includes representatives from Medical Affairs, Regulatory Affairs, Legal, and Marketing. Medical Affairs professionals bring a deep understanding of the clinical data, trial outcomes, and product efficacy, ensuring that promotional claims are scientifically accurate.

Importance for the Business Promotional review is vital for ensuring the company's marketing practices are compliant with legal and regulatory standards. Non-compliance can lead to severe penalties, including fines, legal action, or product withdrawal, which can significantly impact the company's reputation and financial standing. Furthermore, from a business perspective, accurate promotional material builds credibility and trust with HCPs and patients. By ensuring that the messaging is scientifically accurate and balanced, Medical Affairs helps mitigate risks and maintain the integrity of the company's communication with its stakeholders.

In addition to compliance, the participation of Medical Affairs in PRCs adds strategic value by aligning promotional messaging with the latest scientific developments and clinical data. This alignment is crucial when launching new products or expanding indications, as HCPs increasingly rely on transparent, evidence-based information to make prescribing decisions.

Non-Promotional Review Committees

Definition Non-promotional review committees focus on the evaluation of materials that are intended for educational or informational purposes but are not directly related to the promotion of pharmaceutical products. These materials might include medical education content, scientific publications, HCP communications, and responses to unsolicited requests for medical information. Unlike promotional materials, non-promotional content does not advocate for the use of a product but rather provides objective, evidence-based information on diseases, treatment protocols, or product safety.

Purpose The main goal of non-promotional review committees is to ensure that the information disseminated by the company is accurate, unbiased, and scientifically rigorous. This is especially important for medical education initiatives or materials that are intended to foster scientific exchange with HCPs and other stakeholders. Medical Affairs, through its non-promotional review activities, guarantees that such materials are free from marketing influence and aligned with the highest ethical standards.

Non-promotional materials can be critical in educating physicians about emerging scientific trends, new therapeutic approaches, or disease management strategies, all without the intention to promote a specific product. They also serve to build and maintain trust between the company and the scientific community.

Importance for the Business Non-promotional materials, while not designed to increase sales directly, play a vital role in shaping the company's reputation and fostering long-term relationships with HCPs. By disseminating scientifically sound and objective information, Medical Affairs helps establish the company as a leader in clinical expertise and innovation. These efforts can indirectly influence prescribing habits, as HCPs often rely on trusted sources of information when making treatment decisions.

Furthermore, non-promotional content also helps to ensure compliance with industry codes of practice, such as the Pharmaceutical Research and Manufactures of America (PhRMA) Code in the United States or the European Federation of Pharmaceutical Industries and Associations (EFPIA) Code of Practice in Europe, both of which regulate the interaction between the pharmaceutical industry and HCPs. Non-compliance in this area could lead to reputational damage or regulatory penalties.

Medical Affairs' role in reviewing and approving non-promotional content ensures that the company adheres to these guidelines, thus mitigating risk while also providing value through scientific engagement. Importantly, non-promotional activities help to balance the company's communications by offering credible, neutral content that counters any perceptions of bias from promotional campaigns.

Handling Medical Information and Unsolicited Requests for Scientific Information

Definition Medical information refers to the service provided by pharmaceutical companies to respond to inquiries from HCPs, patients, and other stakeholders regarding the company's products. These inquiries are often unsolicited and can cover a wide range of topics, including product safety, efficacy, dosing, off-label use, and more.

Purpose The purpose of handling medical information requests is to provide accurate, balanced, and scientifically grounded responses to queries about the company's products. Medical Affairs is responsible for ensuring that responses are based on the most up-to-date scientific literature and clinical data. Responses must adhere to regulatory standards, which dictate that they should be non-promotional and should not encourage off-label use unless explicitly requested by an HCP in a scientific context.

Importance for the Business From a compliance perspective, the handling of unsolicited requests for medical information is highly regulated. Medical Affairs ensures that responses to such requests are scientifically rigorous and compliant with global and local regulatory frameworks. This is particularly important in cases where the inquiries pertain to off-label use. Providing an inappropriate or promotional response to such queries could lead to significant legal and regulatory repercussions.

The ability to deliver high-quality medical information strengthens the company's relationship with HCPs. Physicians and other healthcare stakeholders often view Medical Affairs as a trusted source of scientific knowledge. By providing clear, evidence-based information, Medical Affairs not only assists in improving patient care but also builds long-term trust and credibility with the scientific community.

The insights gathered from medical inquiries can also be valuable for the company. By analyzing trends in the questions received, Medical Affairs can identify gaps in knowledge, unmet medical needs, or emerging concerns related to the product. These insights can feed back into the company's strategic planning, guiding the development of new educational materials or informing future clinical trials.

Cross-Functional Team Activities (Product Versus Brand Teams)

Definition Cross-functional teams in the pharmaceutical industry consist of members from different departments working together towards a shared objective. These teams can be product-focused (centered on the development or life-cycle management of a particular drug) or brand-focused (focused on the commercial aspects and positioning of a product in the market).

Purpose The purpose of cross-functional teams is to ensure that all perspectives—scientific, commercial, regulatory, and medical—are considered in decision-making processes. For Medical Affairs, this means contributing clinical and scientific insights that help guide product development, regulatory submissions, or commercial strategies. The participation of Medical Affairs in these teams helps ensure that business decisions are aligned with clinical data and that the company maintains a patient-centric approach.

For product teams, Medical Affairs might provide input into clinical trial designs, support safety evaluations, or assist in preparing regulatory submissions. In brand teams, Medical Affairs helps shape the scientific narrative, ensuring that promotional messaging is grounded in the latest clinical data and meets regulatory standards.

Importance for the Business Cross-functional collaboration is essential for ensuring the success of a product throughout its life cycle, from development to post-marketing. Medical Affairs plays a key role in this process by bringing a scientific perspective that informs strategic decisions, whether related to product development, market positioning, or safety monitoring.

In product teams, Medical Affairs ensures that the development of the product aligns with clinical needs and regulatory requirements, helping to de-risk the development process. For example, Medical Affairs can provide insights into patient populations, unmet medical needs, or competitive dynamics, which can guide clinical trial design or label expansion strategies.

In brand teams, Medical Affairs helps ensure that marketing and commercial strategies are grounded in science. By doing so, they prevent the dissemination of information that might be seen as overly promotional or inconsistent with clinical evidence. This balanced approach helps protect the company from regulatory scrutiny and builds long-term trust with HCPs and patients.

By contributing to both product and brand teams, Medical Affairs serves as a bridge between commercial objectives and scientific integrity, ensuring that the company's strategies are both effective and ethical.

The additional enabling activities of Medical Affairs, including participation in promotional and non-promotional review committees, handling medical information requests, and contributing to cross-functional teams, underscore the multifaceted role that Medical Affairs plays in pharmaceutical companies. These activities ensure that business operations are scientifically sound, compliant with regulatory standards, and aligned with patient-centric goals. By participating in these critical functions, Medical Affairs helps to mitigate risks, build trust with stakeholders, and ultimately support the company's strategic objectives across product life cycles.

Measuring the Impact of Medical Affairs

To ensure that the contributions of Medical Affairs are quantifiable, key performance indicators are used to measure the impact of their activities.

Key Performance Indicators (KPIs)

- *Clinical and Commercial Alignment Metrics*

  Metrics that assess the alignment between clinical data and commercial strategy are key indicators of Medical Affairs' success in supporting cross-functional collaboration.
- *Scientific Communication Effectiveness*

  Evaluating the impact of scientific communication, including publications and medical education, helps measure the effectiveness of Medical Affairs in disseminating information.
- *Field Medical Performance*

  MSL activities are often assessed based on their engagements with HCPs and KEEs, as well as the feedback they provide to internal teams.

- *Regulatory Support and Compliance*
  Tracking contributions to regulatory success and compliance activities is another important metric for evaluating Medical Affairs' impact.

Conclusion

Medical Affairs has become a crucial enabling function within the pharmaceutical industry, driving business success while maintaining scientific integrity and ethical standards. Through collaboration with Commercial, Regulatory, and Clinical teams, Medical Affairs enhances product development, market access, and corporate reputation, positioning it as an essential partner in the pharmaceutical ecosystem.

Medical Affairs in Action (Case 1): Impact on a Promotional Review Decision
A large pharmaceutical company was preparing to launch a new biologic treatment for a chronic autoimmune condition. The marketing team developed a promotional campaign that emphasized the drug's superior efficacy compared to current market alternatives. However, during the promotional review process, Medical Affairs identified inconsistencies between the promotional messaging and the clinical data presented in the Phase III trials. Specifically, the marketing materials overstated the significance of the efficacy results in a particular subpopulation.

Medical Affairs Intervention

Medical Affairs flagged this issue during the PRC meeting. The team provided detailed analyses of the clinical data, showing that the statistically significant differences were confined to a specific patient subgroup and not broadly applicable to the entire population as the marketing materials suggested. Medical Affairs highlighted the risk of regulatory scrutiny from agencies like the FDA or EMA if the claims were perceived as misleading or over-exaggerated.

Outcome

The PRC, based on Medical Affairs' input, made substantial revisions to the promotional campaign, ensuring that the messaging was more aligned with the approved clinical data. As a result, the company avoided potential regulatory penalties and the risk of having to retract marketing materials post-launch. The revised campaign was approved and launched without issue, preserving both the company's credibility and the integrity of the product's positioning in the market.

Benefit to Business

This example underscores the critical role of Medical Affairs in safeguarding the company against regulatory risks while ensuring that promotional campaigns are scientifically accurate. Medical Affairs' input helped avoid costly revisions post-launch and ensured compliance with global marketing standards, thereby protecting the company's reputation and fostering trust with both HCPs and regulators.

Medical Affairs in Action (Case 2): Handling Unsolicited Requests for Off-Label Use

A mid-sized pharmaceutical company launched a novel oncology therapy for a specific subtype of breast cancer. Soon after the launch, the company's Medical Information team began receiving unsolicited inquiries from oncologists regarding the potential use of the drug for other cancer subtypes, particularly in patients for whom standard therapies had failed. These off-label requests were outside the approved indication, and handling them improperly could lead to significant regulatory and legal challenges.

Medical Affairs Intervention

The Medical Information team, part of Medical Affairs, responded by carefully crafting responses to each inquiry, ensuring that they adhered to regulatory guidelines that prohibit the promotion of off-label use. The responses were non-promotional, providing balanced, evidence-based information from available clinical trials and RWE. Medical Affairs also emphasized that the data did not currently support the use of the therapy in these unapproved indications but advised physicians to consult the ongoing clinical trials in related areas.

Outcome

By addressing these unsolicited requests effectively, the company avoided regulatory issues related to off-label promotion while still providing HCPs with accurate, evidence-based information. Additionally, the trend of these requests provided valuable insights to the company, which were fed back into the R&D pipeline. Based on the demand and scientific rationale, the company initiated new clinical trials to explore the potential use of the therapy in these other cancer subtypes.

Benefit to Business

Medical Affairs' handling of these unsolicited requests not only protected the company from regulatory scrutiny but also identified new opportunities for product development and label expansion. The feedback from HCPs, collected by the Medical Information team, directly influenced the company's strategic decision-making, demonstrating how Medical Affairs can drive innovation and business growth through its enabling activities.

Medical Affairs in Action (Case 3): Collaboration with Market Access: Supporting Payer Negotiations

In today's competitive pharmaceutical landscape, demonstrating the clinical and economic value of a product is crucial for gaining favorable reimbursement status from payers. Market Access teams rely heavily on robust clinical evidence to support their negotiations with health insurers, government bodies, and other payers. Medical Affairs plays a key role in generating this evidence and translating it into health economics data, which is critical for demonstrating a drug's value proposition.

A company was launching a new orphan drug for a rare genetic disorder. The Market Access team needed to present compelling evidence to justify the high cost of the treatment. Medical Affairs worked closely with the team to compile RWE from post-market surveillance and patient registries, alongside data from clinical trials. This collaboration led to the creation of a comprehensive health economics dossier that illustrated the drug's long-term benefits, including its potential to reduce the need for more expensive treatments and hospitalizations over time.

Outcome and Business Benefit

Thanks to Medical Affairs' contribution, the Market Access team successfully secured reimbursement in key markets. This collaboration helped maximize the product's market penetration and ensured that patients could access the therapy, driving significant revenue growth for the company. This example highlights how Medical Affairs' cross-functional collaboration directly influences market access and commercial success.

Medical Affairs in Action (Case 4): Collaboration with Pharmacovigilance: Ensuring Patient Safety

Pharmacovigilance teams are responsible for monitoring the safety of pharmaceutical products once they are on the market, collecting adverse event reports, and ensuring compliance with regulatory safety standards. Medical Affairs partners with Pharmacovigilance to ensure that safety signals are interpreted within the broader clinical context and communicated effectively to both internal teams and external stakeholders.

Following the launch of an immunotherapy drug, Medical Affairs collaborated with Pharmacovigilance to address reports of a rare but severe immune-related adverse event. Medical Affairs provided insights into the drug's mechanism of action, worked with clinical experts to analyze the cases, and recommended specific updates to safety protocols. They also collaborated with Regulatory Affairs to ensure timely reporting to the FDA and EMA.

Outcome and Business Benefit

By working together, Medical Affairs and Pharmacovigilance were able to mitigate the risks associated with the adverse event, preventing a potential product recall. The company demonstrated its commitment to patient safety, which reinforced trust with HCPs and regulatory bodies. This collaboration helped maintain the product's market position while ensuring ongoing compliance with safety regulations.

Medical Affairs in Action (Case 5): Metrics and KPIs: Quantifying the Impact of Medical Affairs

Measuring the impact of Medical Affairs is critical for demonstrating its value to the business. By tracking KPIs related to promotional review, medical information, and cross-functional collaboration, companies can assess how Medical Affairs' contributions translate into business outcomes.

Promotional Review Metrics

- *Number of Promotional Materials Reviewed*: Tracking the volume of promotional materials reviewed by Medical Affairs can provide insights into its workload and contribution to compliance.

- *Compliance Rates*: Measuring the percentage of materials that are approved without requiring revisions post-launch demonstrates the accuracy and thoroughness of the review process.
- *Time to Review*: Assessing how quickly Medical Affairs can complete the promotional review process helps ensure that marketing timelines are met.

Business Benefit: High compliance rates and timely reviews prevent regulatory fines, protect the company's reputation, and ensure that promotional campaigns are scientifically sound, supporting the successful launch of products.

Medical Information Metrics

- *Average Response Time*: Measuring the average time taken to respond to medical inquiries can indicate how efficiently Medical Affairs handles these requests.
- *Number of Inquiries Handled*: Tracking the volume of medical information inquiries provides insights into the demand for scientific information and potential knowledge gaps among HCPs.
- *Customer Satisfaction Ratings*: Surveying HCPs to gauge their satisfaction with the quality and clarity of responses from Medical Affairs can help improve service levels.

Business Benefit: Faster response times and high satisfaction ratings build trust with HCPs, enhancing the company's reputation and ensuring continued engagement with key stakeholders.

Cross-Functional Collaboration Metrics

- *Impact on Clinical Trial Design*: Measuring the extent to which insights from Medical Affairs influence the design of clinical trials (e.g., identifying relevant endpoints or patient populations) helps demonstrate the team's strategic value.
- *Regulatory Submissions Supported*: Tracking the number of regulatory submissions that Medical Affairs has contributed to, particularly in the form of clinical data and safety evidence, underscores the team's role in regulatory success.
- *Health Economics Data Generated*: Quantifying the number of health economics analyses Medical Affairs contributes to payer submissions demonstrates its influence on market access.

Business Benefit: These metrics highlight how Medical Affairs contributes to clinical development, regulatory success, and market access, all of which are critical to the company's long-term growth and competitiveness.

Cases have provided real-world case studies and expanded on the cross-functional collaboration and KPIs that define the role of Medical Affairs in enabling activities. From influencing promotional review decisions to handling off-label inquiries and collaborating with Market Access and Pharmacovigilance, Medical Affairs adds significant value across the pharmaceutical business landscape. By tracking key metrics and demonstrating contributions to regulatory compliance, patient safety, and market access, Medical Affairs reinforces its role as a strategic partner in driving business success while maintaining scientific integrity.

Chapter 10
Safety and Pharmacovigilance in the Lifecycle of Medicinal Products

Key Highlights

- Pharmacovigilance ensures continuous drug safety monitoring from clinical trials to post-market surveillance.
- Regulatory frameworks (FDA, EMA, WHO) establish global safety monitoring and compliance standards.
- Artificial Intelligence, digital health tools, and Real-wold Evidence enhance adverse event detection and pharmacovigilance efficiency.
- Risk management plans and signal detection prevent and mitigate drug-related risks.
- Transparency, ethical responsibility, and collaboration with regulators ensure patient safety.

Drug safety and pharmacovigilance are fundamental pillars in the pharmaceutical industry, ensuring that medicinal products maintain an acceptable safety profile throughout their lifecycle. These disciplines aim to safeguard public health by systematically monitoring and assessing the adverse effects of drugs and taking proactive measures to mitigate risks. This chapter explores the historical development, core principles, and indispensable role that pharmacovigilance plays from the pre-marketing phase during clinical trials to post-marketing surveillance.

Pharmacovigilance evolved as a response to the recognition that the safety profile of a drug cannot be fully understood through clinical trials alone. Trials, while rigorous, involve controlled environments and relatively small sample sizes compared to the broader patient populations encountered after a drug's release. As a result, pharmacovigilance is essential for identifying and addressing safety issues that may only become apparent during real-world use. This proactive surveillance is particularly vital given the complexities of modern medicines, including biologics and gene therapies, which present unique safety challenges.

A. Krendyukov, *Medical Affairs' Fundamentals: The Next Generation*,
https://doi.org/10.1007/978-3-031-92588-7_10

Historical Development and Evolution of Pharmacovigilance

The roots of modern pharmacovigilance can be traced back to the thalidomide disaster of the 1960s. Thalidomide, initially marketed as a treatment for morning sickness in pregnant women, was later found to cause severe birth defects in thousands of children. This tragic event underscored the critical need for rigorous safety monitoring both before and after a drug enters the market. The thalidomide crisis marked a turning point, leading to the creation of regulatory frameworks for monitoring adverse drug reactions (ADRs) and ensuring that drug safety is prioritized.

Regulatory Frameworks and Global Efforts

In response to this disaster, global regulatory bodies were compelled to establish more structured approaches to drug safety. The World Health Organization (WHO) launched its Programme for International Drug Monitoring in 1968 to facilitate the sharing of drug safety information across countries. This initiative catalyzed the development of national pharmacovigilance systems worldwide.

Over the years, pharmacovigilance has evolved, shaped by emerging safety concerns, advancements in technology, and globalization. Platforms like the International Conference on Harmonisation of Technical Requirements for Pharmaceuticals for Human Use (ICH) have helped harmonize regulatory expectations, ensuring a more coordinated global approach to drug safety. This evolution has been critical in aligning the objectives of drug safety with the rapid pace of pharmaceutical innovation (refer to Table 10.1).

Table 10.1 Key pharmacovigilance (PV) and drug safety acronyms and terms

Acronym	Term	Definition
AE	Adverse event	Any untoward medical occurrence in a patient or clinical trial subject administered a medicinal product, which does not necessarily have a causal relationship with the treatment.
SAE	Serious adverse event	An AE that results in death, is life-threatening, requires hospitalization or prolongation of existing hospitalization, results in persistent or significant disability/incapacity, or is a congenital anomaly/birth defect.
SUSAR	Suspected unexpected serious adverse reaction	A serious adverse reaction that is both unexpected (not consistent with the reference safety information) and suspected to be related to the investigational medicinal product.

(continued)

Table 10.1 (continued)

Acronym	Term	Definition
DSUR	Development safety update report	A periodic report submitted to regulatory authorities summarizing safety information from ongoing clinical trials of an investigational medicinal product.
PSUR (PBRER)	Periodic safety update report (periodic benefit-risk evaluation report)	A comprehensive document providing an evaluation of the benefit-risk balance of an authorized medicinal product at defined time intervals.
RMP	Risk management plan	A detailed document describing a medicinal product's identified and potential risks and how they will be managed during clinical development and post-marketing phases.
ICSR	Individual case safety report	A report detailing a single occurrence of an AE or ADR, containing relevant clinical and patient information, submitted to regulatory authorities.
MedDRA	Medical dictionary for regulatory activities	A standardized medical terminology used globally to classify and code AE and ADR data.
GVP	Good pharmacovigilance practices	A set of measures designed to ensure the safety monitoring and risk management of medicinal products in the European Union.
CIOMS	Council for International Organizations of medical sciences	An organization that provides internationally recognized pharmacovigilance guidelines and reporting standards, such as the CIOMS I form for ADR reporting.
ICH E2E	International Council for Harmonization of technical requirements for Pharmaceuticals for Human use (E2E guidelines)	A set of guidelines that outline pharmacovigilance planning and risk management strategies.
MAH	Marketing authorization holder	The pharmaceutical company that holds the license to market a medicinal product and is responsible for its safety monitoring.
PV	Pharmacovigilance	The science and activities relating to the detection, assessment, understanding, and prevention of adverse effects or any other drug-related problems.
EudraVigilance	European database for pharmacovigilance	A system designed by the European medicines agency (EMA) for managing and analyzing safety reports related to medicinal products approved in the EU.
FDA FAERS	FDA adverse event reporting system	The U.S. FDA'S database for monitoring adverse event reports and medication errors for marketed drugs and biologics.

(continued)

Table 10.1 (continued)

Acronym	Term	Definition
REMS	Risk evaluation and mitigation strategy	A drug safety program required by the FDA to manage known or potential serious risks associated with certain medications.
Signal detection	Safety signal detection	The process of identifying new or known safety concerns related to a medicinal product through data analysis of spontaneous reports, clinical trials, or literature.
PMS	Post-marketing surveillance	The monitoring of a drug's safety after it has been approved and released on the market.
BRR	Benefit-risk ratio	An assessment of the positive therapeutic effects of a drug relative to its potential risks.
CA	Competent authority	The national regulatory agency responsible for overseeing clinical trials, drug approvals, and pharmacovigilance activities (e.g., EMA, FDA, MHRA).
CAPA	Corrective and preventive action	A systematic approach to identifying, correcting, and preventing issues in pharmacovigilance, quality assurance, and regulatory compliance.
CRF	Case report form	A standardized document used in clinical trials to record data on each study participant, including AEs and SAEs.
NISS	Non-interventional safety study	A post-marketing observational study conducted to gather additional safety data without modifying standard medical treatment.
PASS	Post-authorization safety study	A study conducted after a medicinal product is authorized to evaluate its safety in real-world use.
PAES	Post-authorization efficacy study	A study designed to further assess the efficacy of a medicinal product after it has been approved for marketing.
PV system master file (PSMF)	Pharmacovigilance system master file	A regulatory-required document that provides an overview of a company's pharmacovigilance system, including risk management procedures.
RA	Regulatory authority	A governmental body responsible for ensuring the safety, efficacy, and quality of medicinal products (e.g., EMA, FDA, TGA, MHRA).

Importance of Drug Safety and Pharmacovigilance

Guardians of Patient Safety

At the heart of drug safety and pharmacovigilance lies an unwavering commitment to protecting patients. Pharmacovigilance activities—such as collecting, evaluating, and interpreting ADRs—serve as the first line of defense against unforeseen risks. These activities ensure that even after a drug is approved, its safety profile continues to be scrutinized and updated in light of new information.

Pre-marketing and Post-marketing Surveillance

Pharmacovigilance plays a crucial role in both the pre-marketing and post-marketing phases of a drug's lifecycle. During clinical trials, pharmacovigilance helps identify potential safety concerns, allowing for early intervention before a drug is widely distributed. Post-marketing, pharmacovigilance is essential for monitoring the drug's safety in real-world settings, where diverse patient populations and comorbidities may reveal adverse effects that were not evident in controlled trials.

For example, in a clinical trial, a drug may show no severe side effects in a relatively small group of patients. However, when the drug is released into the general population, pharmacovigilance systems are tasked with monitoring its use among a much larger and more diverse group of people. This surveillance is critical for detecting rare but serious ADRs, such as liver toxicity or cardiovascular risks, which may require regulatory action, including label changes, warnings, or, in extreme cases, withdrawal of the product from the market.

Regulatory Framework and Requirements

The global regulatory landscape for pharmacovigilance is shaped by coordinated efforts to standardize drug safety reporting and monitoring across regions. The ICH has been instrumental in this regard, developing guidelines that provide a unified approach to clinical safety data management, including the reporting of ADRs.

ICH's E2E guidelines outline procedures for detecting, assessing, and reporting safety data. Compliance with these guidelines ensures that pharmaceutical companies are able to meet their safety obligations globally, facilitating the smooth conduct of clinical trials and marketing approval processes across different regions.

Despite global harmonization efforts, regional regulatory bodies maintain specific pharmacovigilance requirements to address the unique characteristics of their healthcare environments. For instance, the US Food and Drug Administration (FDA), European Medicines Agency (EMA), and Japan's Pharmaceuticals and

Medical Devices Agency (PMDA) each have distinct reporting timelines and formats for ADR submissions.

Understanding and navigating these regional differences is essential for companies operating globally. For example, while the EMA mandates periodic safety update reports (PSURs) post-approval, the FDA requires specific adverse event reports (AERs) to be submitted under stringent timelines. A failure to meet these requirements could lead to fines, delays in approval, or withdrawal of market authorization.

Pharmacovigilance regulations are not static; they evolve alongside medical and technological advancements. With the rise of biologics, gene therapies, and digital health solutions, regulatory bodies are increasingly focused on adapting their frameworks to address new safety challenges. Emerging fields such as personalized medicine also necessitate a more individualized approach to pharmacovigilance, ensuring that the safety of complex treatments is monitored as rigorously as traditional pharmaceuticals.

Needs Assessment for Establishing a Pharmacovigilance Department

Establishing a robust pharmacovigilance department is crucial for pharmaceutical companies to ensure they can manage the safety of their products effectively. A comprehensive needs assessment is the first step in determining the appropriate structure, resources, and strategies required to support pharmacovigilance activities.

Scale of Operations and Portfolio Complexity

The size and complexity of a company's product portfolio directly influence the structure of its pharmacovigilance department. Companies with extensive pipelines, including biologics and orphan drugs, require specialized pharmacovigilance systems capable of handling complex safety issues. Additionally, companies operating in multiple regions must ensure compliance with diverse regulatory requirements, further underscoring the need for a scalable pharmacovigilance infrastructure.

Integration with Quality Management Systems

Pharmacovigilance should not function in isolation but rather be integrated into the company's broader quality management system. Safety data should be seamlessly shared across departments, including Clinical Development, Regulatory Affairs, and Medical Affairs, to ensure comprehensive safety oversight. This integration is essential for fostering a culture of safety throughout the company and ensuring that safety signals are detected and acted upon swiftly.

Challenges in Implementing Pharmacovigilance

Pharmaceutical companies face numerous challenges in implementing effective pharmacovigilance systems. These challenges span from data collection to signal detection and regulatory compliance, each of which must be addressed to ensure a robust safety monitoring process.

Data Collection and Signal Detection

One of the most significant challenges is the accurate and comprehensive collection of safety data. Underreporting of ADRs remains a widespread issue, particularly in post-marketing settings where healthcare professionals may not always report minor or expected adverse events. Incomplete or inconsistent data also complicates signal detection, making it difficult to differentiate between normal background noise and genuine safety signals.

Global Reporting and Compliance

Navigating the global regulatory landscape adds another layer of complexity. Each region has its own reporting requirements, timelines, and formats, requiring companies to develop tailored pharmacovigilance strategies for each market. Additionally, as regulations evolve, companies must stay abreast of changes to ensure they remain compliant or risk facing penalties.

Opportunities and Innovations in Pharmacovigilance

While pharmacovigilance faces many challenges, technological advancements and collaborative efforts are creating new opportunities for improving the efficiency and effectiveness of safety monitoring.

Artificial intelligence (AI) and big data analytics are revolutionizing pharmacovigilance by automating the detection of safety signals from vast datasets. Machine learning algorithms can sift through electronic health records (EHRs), social media, and other real-world data sources to identify patterns that may indicate adverse drug reactions (ADRs). These technologies offer significant promise for early signal detection and can complement traditional pharmacovigilance methods.

Digital health technologies are empowering patients to play a more active role in pharmacovigilance. Mobile applications and online portals allow patients to report adverse events in real-time, providing pharmacovigilance teams with more immediate access to safety data. These technologies also foster patient engagement, improving the quality and quantity of data collected in real-world settings.

International Collaboration and Best Practices

The global nature of pharmacovigilance demands international collaboration to ensure that safety data are shared efficiently and that best practices are adopted across regions.

Collaborative platforms, such as those established by the WHO and ICH, facilitate the sharing of safety data across regions. These platforms allow for the pooling of resources and expertise, enabling faster identification and resolution of global safety issues. Participation in these initiatives is critical for companies looking to strengthen their pharmacovigilance systems and improve patient safety worldwide.

Expanding Product Safety and Pharmacovigilance in the Modern Era

As drug development becomes more complex and patient populations more diverse, pharmacovigilance must evolve to meet these challenges. By leveraging cutting-edge technologies, embracing transparency, and maintaining a patient-centered approach, the pharmaceutical industry can improve drug safety and enhance patient outcomes.

Risk Management Plans (RMPs): Essential Tools in Pharmacovigilance

RMPs are structured strategies designed to identify, characterize, and minimize risks associated with medicinal products. They are a core component of pharmacovigilance, ensuring that any potential risks are adequately managed throughout a drug's lifecycle, from pre-market approval through post-marketing surveillance.

Components of RMPs

RMPs typically include the following components:

- *Safety Specification*: A detailed overview of identified and potential risks, including important missing information.
- *Pharmacovigilance Plan*: This outlines activities aimed at identifying and characterizing safety concerns. It may include studies and clinical trials designed to collect additional safety data.
- *Risk Minimization Measures (RMMs)*: Strategies to reduce the occurrence or severity of adverse effects, including healthcare provider training, labeling changes, or patient education programs.
- *Benefit-Risk Evaluation*: A continuous assessment of whether the benefits of the drug outweigh the risks.

These components ensure that any emerging risks are quickly identified and appropriately managed, with mitigation strategies implemented as needed.

Integration with Clinical Trials

RMPs significantly influence the design of clinical trials, particularly Phase III and post-approval Phase IV studies. During clinical trials, RMPs help to focus safety data collection on the identified risks and guide the decision-making process when it comes to patient monitoring, study design, and endpoints. For example, if a drug is known to cause hepatotoxicity, the clinical trial may include more frequent liver function tests and require long-term follow-up.

Signal Detection in the Digital Era: AI and New Technologies

In modern pharmacovigilance, signal detection refers to the process of identifying new or unknown safety risks based on data from a variety of sources, including clinical trials, spontaneous reporting systems, and RWE. In the digital age, advancements in technology are revolutionizing how safety signals are detected and acted upon.

AI in Signal Detection

AI and machine learning (ML) have transformed pharmacovigilance by automating the detection of safety signals. These technologies can process vast datasets from diverse sources, such as EHR, social media, and clinical trial data, to identify patterns that may indicate emerging safety concerns.

- *Predictive Analytics*: AI-driven algorithms can predict potential safety issues based on patient characteristics, drug interactions, or genetic factors. This proactive approach helps prevent adverse events before they occur.

Early Signal Detection: AI systems can detect signals earlier than traditional methods by continuously monitoring real-world data in real-time, allowing faster identification of rare or unexpected adverse events.

Social Media and Wearables: New Frontiers for Signal Detection

In addition to AI, social media platforms and wearable devices have emerged as valuable sources of real-world data. Social media provides a wealth of unstructured data on patient experiences with medications, while wearable devices offer continuous monitoring of vital signs and other health metrics.

- *Social Media Mining*: Patients often share their experiences with medications on social media, including any side effects they encounter. Advanced text mining tools can analyze these posts to identify potential safety signals.

Wearables and EHRs: Real-world data from wearable health devices, combined with EHRs, can provide real-time insights into how drugs affect patients' daily lives. For example, a sudden change in heart rate detected by a wearable could signal a cardiovascular event related to a drug.

Global Pharmacovigilance Databases: Enhancing Collaboration

The global nature of the pharmaceutical industry necessitates the sharing of safety data across borders. International pharmacovigilance databases such as EudraVigilance (in the EU) and VigiBase (run by the WHO) play a critical role in enhancing transparency and facilitating the early detection of safety signals on a global scale.

EudraVigilance

EudraVigilance is a centralized database used by the EMA for managing and analyzing information on suspected adverse reactions to medicines authorized in the European Economic Area (EEA). It enables real-time monitoring and prompt responses to safety issues.

- *Signal Management*: EudraVigilance has sophisticated signal management tools that allow regulatory authorities to assess and prioritize potential safety concerns. This helps in swift decision-making, leading to quicker implementation of risk minimization measures when needed.

VigiBase

VigiBase, operated by the WHO's Uppsala Monitoring Centre, is the largest global pharmacovigilance database. It collects ADR reports from over 130 countries, facilitating the identification of emerging safety issues worldwide.

- *Example:* A notable success of VigiBase was its role in the early detection of safety signals for the antimalarial drug chloroquine, which led to concerns over potential cardiotoxicity in certain patient populations. Early identification enabled timely regulatory action, preventing widespread harm.

Challenges of Global Databases

While these global pharmacovigilance databases enhance safety monitoring, they face challenges, particularly in harmonizing data formats and definitions across countries. Differences in reporting standards and the quality of data submitted can complicate signal detection and risk assessments.

The Role of RWE in Pharmacovigilance

RWE is increasingly recognized as a critical tool for pharmacovigilance. Unlike data from controlled clinical trials, RWE is derived from real-world settings, where a drug is used by patient populations with various co-morbidities and medications.

RWE complements clinical trials by providing insights into how a drug performs in broader, more diverse populations. It helps to identify long-term adverse effects or interactions that may not have been evident during controlled trials. This integration ensures a more comprehensive understanding of a drug's safety profile.

In the post-marketing phase, RWE becomes essential for monitoring drug safety in real-world settings. For instance, data from insurance claims, patient registries, or hospital records can highlight trends in adverse drug reactions over time.

Regulatory Implications of RWE

Regulatory bodies are increasingly incorporating RWE into their decision-making processes. For instance, the FDA's *Real-World Evidence Program* encourages the use of RWE for regulatory purposes, including drug approvals for new indications, safety monitoring, and post-marketing commitments.

Ethical Considerations in Pharmacovigilance

Ethical considerations are a cornerstone of pharmacovigilance, particularly as the complexity of drug safety monitoring increases.

Ethical Implications of Delayed Reporting

The consequences of delaying the communication of adverse events can be severe, as seen at the end of this chapter in one of the case studies. Ethical responsibility dictates that pharmaceutical companies must report emerging risks promptly to regulatory authorities and the public. Delayed reporting not only endangers patients but also erodes public trust in the healthcare system.

Patient Rights and Informed Consent

Pharmacovigilance data should inform patient consent during clinical trials. Patients have the right to be fully informed of the risks associated with a drug, especially as new safety data emerges. Clear and transparent communication about risks is not just a regulatory requirement but an ethical obligation.

Transparency in Data Sharing

Pharmaceutical companies have an ethical duty to share safety data transparently, both with regulatory authorities and with HCPs. This transparency ensures that healthcare professionals can make informed decisions when prescribing medications and that patients receive accurate information about the risks and benefits of their treatments.

The Role of Qualified Persons for Pharmacovigilance (QPPVs)

QPPVs are responsible for overseeing pharmacovigilance activities within pharmaceutical companies. Their role is critical in ensuring compliance with regulatory requirements and maintaining the safety of medicinal products.

Responsibilities of QPPVs

- Ensuring that pharmacovigilance systems are in place and function effectively.
- Overseeing the collection, analysis, and reporting of adverse events.
- Communicating with regulatory authorities regarding any emerging safety concerns.

Challenges Faced by QPPVs

QPPVs often face challenges in managing pharmacovigilance activities across multiple regions with differing regulatory requirements. Ensuring consistency and compliance across global operations requires robust systems and effective communication.

Cross-Departmental Integration in Pharmacovigilance

For pharmacovigilance to be effective, it must be integrated across various departments within a pharmaceutical company, including Clinical Research, Regulatory Affairs, and Medical Affairs.

Clear communication channels are essential for sharing safety data in real time. Pharmacovigilance teams must work closely with clinical research teams to ensure that any safety signals identified during trials are communicated promptly. This collaboration helps in refining clinical trial designs and enhancing patient monitoring.

Many companies are investing in integrated technology platforms that allow seamless access to pharmacovigilance data across departments. These systems enable better collaboration, faster decision-making, and a more comprehensive approach to drug safety monitoring.

Future Trends in Pharmacovigilance Regulation

The future of pharmacovigilance is moving toward more real-time safety monitoring and personalized approaches to drug safety, particularly as precision medicine continues to evolve.

Regulatory bodies are increasingly focused on incorporating patient-reported outcomes, big data, and RWE into pharmacovigilance practices. As new therapeutic approaches, such as gene therapies and personalized medicines, become more prevalent, regulators are adapting their frameworks to ensure these complex products are monitored effectively.

Precision medicine, which tailors treatments to individual patients based on their genetic profiles, requires a more personalized approach to safety monitoring. Pharmacovigilance systems will need to evolve to track and manage safety risks for smaller patient populations with highly individualized treatments.

Continuous Vigilance: A Call to Action

The complexity of medicinal products continues to grow, making continuous pharmacovigilance more important than ever. Pharmaceutical companies, regulators, and HCPs must collaborate to enhance safety monitoring and ensure that patients remain protected from unforeseen risks.

Embrace New Technologies

By adopting innovative technologies such as AI, real-world data analytics, and integrated platforms, pharmacovigilance teams can enhance their ability to detect and mitigate risks early.

Prioritize Patient Safety

Patient safety must always be the top priority. Pharmaceutical companies must commit to transparency and ethical practices, ensuring that risks are communicated promptly and that patient well-being is prioritized over commercial interests.

Global Collaboration

International collaboration is critical for addressing global safety issues and sharing best practices. As the pharmaceutical industry becomes increasingly interconnected, global efforts are needed to ensure a consistent, effective approach to pharmacovigilance.

Summary

Pharmacovigilance is an essential, dynamic component of drug safety that extends across the entire lifecycle of medicinal products. As pharmaceutical innovation continues to advance, so too must pharmacovigilance practices evolve. Through international collaboration, the adoption of innovative technologies, and a commitment to patient-centered care, the pharmaceutical industry can continue to improve the safety and effectiveness of drugs. By embracing these changes, stakeholders—regulators, sponsors, and HCPs—can collectively contribute to a safer, more resilient healthcare landscape, fulfilling their shared responsibility to protect public health.

The evolving landscape of pharmacovigilance offers both challenges and opportunities. By leveraging modern technologies, adhering to ethical standards, and fostering collaboration across departments and borders, the pharmaceutical industry can continue to improve product safety and protect patient health.

Case Study: Product Safety and Pharmacovigilance
In 2016, a global pharmaceutical company developed and launched a biologic medicine aimed at treating moderate to severe rheumatoid arthritis. The product was a monoclonal antibody that targeted a specific inflammatory cytokine pathway, offering an innovative approach to reducing inflammation and improving patient outcomes. After an extensive development process, including successful Phase III clinical trials, the product was approved by major regulatory agencies, such as the FDA and EMA, and quickly became available in multiple global markets.

Initially, the product's safety profile appeared promising, with Phase III trials showing manageable ADRs consistent with other immunosuppressive therapies. Common side effects like mild infections and injection site reactions were anticipated and clearly communicated in the labeling.

However, within the first year of post-marketing surveillance, a concerning trend began to emerge. Numerous reports of serious adverse events involving opportunistic infections and severe allergic reactions were submitted to regulatory bodies and the company's pharmacovigilance department. Among these reports, there was an alarming rise in cases of progressive multifocal leukoencephalopathy (PML), a rare and often fatal brain infection associated with immunosuppression.

Root Cause Analysis

The company's pharmacovigilance department, using their safety monitoring systems, flagged these cases for further investigation. A dedicated team was formed, including pharmacovigilance specialists, safety data analysts, clinical researchers, and regulatory experts, to conduct a thorough analysis of the reported cases. The objectives were to understand the underlying causes of the PML cases and determine whether there was a causal relationship between the drug and the emerging safety signals.

Data Review and Signal Detection

The initial step in the root cause analysis was to collect and review all reported cases of PML from various sources, including spontaneous reports from healthcare professionals, patient-reported events through a dedicated portal, and data from existing patient registries. The company also collaborated with regulatory authorities to gather additional reports from their respective safety databases, including the FDA Adverse Event Reporting System (FAERS) and EMA's EudraVigilance system.

Upon review, the pharmacovigilance team observed that the majority of PML cases occurred in patients who had been on the biologic therapy for more than 6 months and had received concomitant treatments with other immunosuppressants, particularly corticosteroids. Additionally, it was noted that many of the affected patients had underlying comorbidities that increased their susceptibility to infections, such as HIV or previous organ transplants.

Hypothesis Generation and Biological Plausibility

The next phase of the analysis focused on determining the biological plausibility of the drug being associated with the increased risk of PML. Given that PML is caused by the reactivation of the John Cunningham (JC) virus, typically in immunocompromised individuals, the pharmacovigilance team hypothesized that the biologic drug, by targeting key inflammatory pathways, might be increasing the vulnerability of patients to viral reactivation.

This hypothesis was supported by preclinical studies that suggested the drug could potentially disrupt immune surveillance mechanisms that normally keep latent viral infections, such as the JC virus, in check. Additionally, pharmacokinetic and pharmacodynamic data from the clinical trials were revisited to understand the drug's long-term immunosuppressive effects, particularly in combination with other immunosuppressive therapies.

Comparative Analysis with Similar Biologics

To gain further insights, the team conducted a comparative analysis of safety data from similar biologics targeting the same or related cytokine pathways. The analysis revealed that while other biologics had also been associated with opportunistic infections, the frequency and severity of PML cases with this specific product were notably higher. This finding prompted further scrutiny of the product's mechanism of action and its potential to amplify immunosuppressive effects in certain patient populations.

Risk Minimization and Regulatory Response

After the root cause analysis established a probable link between the biologic medicine and an elevated risk of PML, particularly in patients with additional risk factors, the company moved swiftly to implement risk minimization strategies. These actions were taken in collaboration with regulatory authorities to ensure that patient safety was prioritized and that the product remained available for those who benefited from its therapeutic effects.

The first step was to update the medicine's labeling to reflect the newly identified risks. The company added a black box warning—the FDA's most serious type of warning—on the increased risk of PML, specifically highlighting the danger in patients who were concurrently using other immunosuppressive therapies or had underlying conditions that compromised their immune systems. The updated label also provided detailed guidance on screening patients for risk factors before initiating therapy and recommended regular monitoring for signs of infection during treatment.

Additionally, the labeling changes included enhanced instructions on discontinuing the drug in patients who developed signs of JC virus reactivation or any neurological symptoms suggestive of PML.

Recognizing the critical role HCPs play in identifying and managing ADRs, the company launched an educational campaign targeted at physicians, nurses, and pharmacists. This campaign included continuing medical education (CME) courses, webinars, and printed materials that emphasized the importance of patient screening, risk factor assessment, and ongoing monitoring.

The company also collaborated with professional societies in rheumatology and immunology to disseminate the updated safety information and provide clinical guidance on managing patients who were already on the drug. These efforts aimed to equip healthcare professionals with the tools and knowledge needed to make informed decisions about patient selection and management, thereby reducing the risk of serious infections.

Risk Evaluation and Mitigation Strategy (REMS)

In response to the emerging safety concerns, the FDA mandated a Risk Evaluation and Mitigation Strategy (REMS) for the biological product. The REMS required the company to implement specific measures to ensure that the benefits of the drug outweighed its risks. These measures included restricted distribution of the drug, where only certified prescribers who had completed risk management training were allowed to prescribe the medication.

The REMS program also mandated the creation of a patient registry to track long-term outcomes and monitor for additional cases of PML. This registry served as a critical tool for gathering RWE, allowing the company to continuously assess the drug's safety profile and make adjustments to the risk management plan as needed.

Lessons Learned

- *Importance of Continuous Post-Marketing Surveillance*

One of the most important lessons learned from this case was the critical role of post-marketing surveillance in ensuring the long-term safety of biologic drugs. Despite rigorous clinical trials, certain ADRs may not become evident until the drug is used in a broader patient population, especially for drugs that modulate the immune system. Continuous safety monitoring allows for the early detection of rare but serious events, enabling timely interventions to protect patient health.

This case highlighted the need for robust pharmacovigilance systems that can detect safety signals early and conduct thorough analyses to determine the root causes of emerging issues. The use of advanced data analytics, combined with traditional safety monitoring practices, proved essential in uncovering the connection between the biologic drug and PML.

- *Proactive Collaboration with Regulatory Authorities*

Proactive collaboration with regulatory authorities, such as the FDA and EMA, was crucial in addressing the safety concerns swiftly and transparently. By working closely with these agencies, the company was able to update labeling, implement REMS programs, and launch educational initiatives in a coordinated manner. This ensured that both HCPs and patients were informed about the risks and could take appropriate actions to mitigate them.

The case underscored the importance of maintaining open lines of communication with regulatory agencies, not only to comply with legal requirements but also to foster a spirit of shared responsibility in safeguarding public health.

- *Tailored Risk Management for High-Risk Patient Populations*

Another key lesson was the importance of tailoring risk management strategies to specific patient populations. In this case, the risk of PML was heightened in patients with certain risk factors, such as concomitant immunosuppressive therapy or underlying immunodeficiencies. Identifying these subpopulations and implementing targeted risk minimization strategies, such as enhanced screening and regular monitoring, was critical in reducing the incidence of serious adverse events.

This approach also emphasized the value of personalized medicine, where individual patient characteristics are considered in the decision-making process for initiating and continuing treatment. Pharmacovigilance plays a crucial role in this personalized approach by providing the data needed to understand how different patient groups respond to therapies.

- *The Role of Patient Engagement in Pharmacovigilance*

Finally, the case illustrated the growing importance of patient engagement in pharmacovigilance. The availability of online portals and mobile applications for reporting ADRs empowered patients to take an active role in monitoring their health and reporting potential side effects. This influx of real-world data from patients provided valuable insights that complemented traditional reporting mechanisms, leading to a more comprehensive understanding of the drug's safety profile.

Moving forward, fostering patient engagement through digital health technologies and patient education will be essential in enhancing the effectiveness of pharmacovigilance systems. By actively involving patients in safety monitoring, pharmaceutical companies can improve data collection and ensure that potential risks are identified and addressed promptly.

This case highlights the dynamic and complex nature of product safety and pharmacovigilance, particularly in the context of biologic drugs. The discovery of serious safety signals, such as PML, underscores the importance of continuous post-marketing surveillance and the need for collaboration between pharmaceutical companies, HCPs, and regulatory authorities. By implementing proactive risk management strategies and tailoring interventions to high-risk patient populations, pharmaceutical companies can mitigate safety risks while ensuring that patients continue to benefit from innovative therapies.

Case study: Product "R"—A Landmark in Pharmacovigilance

Product R, a COX-2-selective non-steroidal anti-inflammatory drug (NSAID), was initially hailed as a breakthrough treatment for pain and inflammation, particularly in patients suffering from osteoarthritis and rheumatoid arthritis. Approved by the FDA in 1999, Product R quickly gained popularity due to its ability to reduce gastrointestinal side effects commonly associated with other NSAIDs. The drug generated significant revenue and became one of its manufacturer's flagship products.

However, less than 5 years after its market launch, reports began emerging regarding potential cardiovascular risks associated with long-term use of Product R. In particular, researchers observed that patients taking Product R for extended periods appeared to have an increased risk of heart attacks and strokes. The discovery of these adverse effects raised alarm within the medical community and triggered closer scrutiny of the drug's safety profile.

Early Observations:

- In the years following its approval, a number of post-marketing reports and independent studies began to signal elevated rates of cardiovascular events among patients on long-term Product R therapy.
- The company conducted its own clinical trial, VIGOR, to assess gastrointestinal safety. While the study confirmed that Product R reduced gastrointestinal issues, it unexpectedly revealed an increased incidence of cardiovascular events in patients using Product R compared to those on naproxen, another NSAID.
- This discrepancy was initially downplayed, with the company suggesting that naproxen might have cardioprotective effects rather than recognizing Product R as a contributor to cardiovascular harm.

Despite these warning signs, the magnitude of the problem did not fully come to light until 2004, when the company's APPROVe trial—designed to study Product R's ability to prevent colon polyps—demonstrated a clear link between long-term Product R use and an increased risk of heart attack and stroke. Based on this evidence, the company voluntarily withdrew Product R from the market in September 2004, marking one of the largest drug withdrawals in history.

Root Cause Analysis: What Went Wrong?

The Product R case highlighted significant gaps in pharmacovigilance and risk management strategies, particularly with regard to long-term drug safety monitoring. A detailed analysis of the situation revealed multiple root causes:

- *Inadequate Post-Marketing Surveillance*: Despite early warning signs from the VIGOR trial and post-market reports, the company continued to market Product R aggressively without adequately addressing the emerging safety concerns. The post-marketing surveillance system failed to promptly capture and interpret cardiovascular safety signals, delaying critical interventions.
- *Flawed Clinical Trial Design*: The VIGOR trial was designed to assess gastrointestinal outcomes but failed to prioritize the monitoring of cardiovascular risks, which were outside the primary scope of the study. Additionally, the trial excluded patients who were at higher cardiovascular risk, limiting the ability to detect significant safety issues in a broader population.
- *Selective Interpretation of Data*: The company initially downplayed the cardiovascular risks observed in VIGOR by suggesting that naproxen's cardioprotective effects might explain the differences. This selective interpretation of data contributed to a delay in acknowledging the true risks associated with Product R.
- *Regulatory Gaps and Communication*: Although the FDA was aware of the cardiovascular signals as early as 2001, there were delays in taking regulatory action, such as requiring stronger warnings on Product R's labeling. There appeared to be a lack of timely and transparent communication between the pharmaceutical company, regulatory agencies, HCPs, and the public about the potential risks.
- *Overreliance on Short-Term Data*: Product R had been approved based on clinical trials that primarily focused on short-term outcomes. The long-term safety profile of the drug, particularly concerning cardiovascular health, was not fully understood at the time of approval, underscoring the importance of continuous monitoring beyond the initial regulatory approval.

Regulatory and Industry Response.

In response to the Product R crisis, significant regulatory and industry-wide changes were implemented to strengthen pharmacovigilance and risk management practices. These changes were intended to prevent similar occurrences and ensure that patient safety remained at the forefront of drug development and post-marketing activities.

- *Strengthened Post-Marketing Surveillance Systems*: Following the Product R withdrawal, regulatory agencies worldwide enhanced their post-marketing surveillance systems, placing greater emphasis on the detection of long-term safety issues. The FDA introduced the REMS program, requiring pharmaceutical companies to implement risk management plans that proactively monitor drug safety after approval.

Additionally, the FDA's Sentinel Initiative, launched in 2008, aimed to create an active surveillance system by leveraging large databases of healthcare information to detect early safety signals in real-world settings.

- *Mandatory Post-Marketing Studies*: Pharmaceutical companies are now required to conduct post-marketing (Phase IV) studies to gather additional data on the safety of their products over extended periods. These studies are often designed to capture long-term adverse effects, such as cardiovascular risks, which may not be fully apparent during pre-approval trials.
- *Improved Labeling and Risk Communication*: In response to the Product R case, regulators began implementing stricter requirements for drug labeling. Manufacturers must provide clear and transparent information on potential risks, including warnings about known adverse events, in both pre-marketing and post-marketing stages. Product R's initial label did not adequately communicate the cardiovascular risks, which became a focal point in the reform of risk communication protocols.
- *Global Harmonization of Pharmacovigilance Guidelines*: The ICH and other global regulatory bodies worked toward the harmonization of pharmacovigilance standards. The ICH guidelines, particularly those related to clinical safety data management and risk evaluation, became instrumental in ensuring that pharmaceutical companies applied consistent safety monitoring practices across different regions.
- *Emphasis on Transparency and Data Sharing*: The Product R case underscored the importance of transparency in clinical trial data reporting. Regulators and industry stakeholders pushed for greater access to both pre- and post-approval data, encouraging pharmaceutical companies to share safety information openly with regulators, healthcare professionals, and the public.

Lessons Learned: Key Takeaways from the Product R Case

The Product R case served as a watershed moment in the history of drug safety and pharmacovigilance, highlighting the need for continuous improvement in both regulatory oversight and industry practices. Several critical lessons emerged from this case:

- *Importance of Long-Term Safety Data*: The absence of comprehensive long-term data on Product R's cardiovascular risks contributed to the delayed identification of serious adverse effects. Pharmaceutical companies must prioritize the collection of long-term safety data, even after regulatory approval, to ensure the ongoing safety of medicinal products.
- *Proactive Risk Management*: Pharmaceutical companies must adopt a proactive approach to pharmacovigilance, identifying and mitigating risks before they become major public health concerns. Relying on short-term clinical trial data is insufficient, particularly for drugs intended for chronic use.
- *Transparency and Ethical Responsibility*: The selective interpretation of safety data and the lack of timely disclosure about potential cardiovascular risks were critical failings in the Product R case. Ethical responsibility demands that

companies be transparent about safety concerns and communicate risks openly to regulators, HCPs, and patients.

- *Need for Robust Post-Marketing Surveillance*: The Product R crisis underscored the inadequacy of traditional passive surveillance systems in identifying safety signals. Active surveillance systems, such as the FDA's Sentinel Initiative, are necessary to detect adverse events in real-world settings more effectively.
- *Enhanced Regulatory Collaboration*: A coordinated global response to pharmacovigilance is essential in a world where pharmaceutical products are distributed across multiple regions. Harmonizing safety monitoring practices, data collection, and risk communication across regulatory bodies ensures more efficient responses to safety concerns.

The withdrawal of Product R from the market had far-reaching consequences for the pharmaceutical industry and regulatory bodies alike. It acted as a wake-up call, compelling stakeholders to rethink and reform how drug safety is monitored both during clinical trials and after a product's release. The case demonstrated that even high-profile drugs could harbor unforeseen risks, reinforcing the importance of pharmacovigilance as a continuous, lifecycle-spanning responsibility.

Moving forward, the lessons learned from Product R have shaped a more vigilant, transparent, and patient-centric approach to pharmacovigilance. The implementation of stronger post-marketing surveillance systems, enhanced regulatory frameworks, and a commitment to long-term safety monitoring ensures that the Product R experience remains a cornerstone in the ongoing evolution of drug safety.

Chapter 11
Medical Affairs and Real-World Evidence

Key Highlights

- RWE enhances understanding of treatment effectiveness, safety, and utilization in real-world settings.
- Differentiation between RWE, RWD, and randomized controlled trials (RCTs) is essential for appropriate data application.
- HEOR complements RWE by incorporating economic analyses into treatment evaluations.
- Key data sources include EHRs, patient registries, claims databases, and PROs.
- AI and digital health tools optimize RWE collection, analysis, and regulatory applications.

In the evolving landscape of healthcare, the demand for data-driven insights has never been greater. Real-world evidence (RWE) has emerged as a pivotal tool in the pharmaceutical industry, offering companies the ability to understand how therapies perform outside of controlled clinical trial settings. Medical Affairs plays a crucial role in the collection, interpretation, and application of RWE, ensuring that pharmaceutical decisions are grounded in evidence that reflects actual patient experiences. This chapter will explore the definition of RWE and how it differs from related concepts such as real-world data (RWD) and randomized controlled trials (RCTs), as well as its intersection with health economics and outcomes research (HEOR). It will also discuss the methodologies used to gather and analyze RWE and the essential role of Medical Affairs in shaping business strategies through the use of RWE.

A. Krendyukov, *Medical Affairs' Fundamentals: The Next Generation*,
https://doi.org/10.1007/978-3-031-92588-7_11

Defining Real-World Evidence

Real-world evidence refers to clinical evidence that is derived from the analysis of RWD—information gathered from everyday healthcare settings rather than controlled clinical trials. RWE provides insights into how a medical product performs in broader and more diverse patient populations, offering a more holistic view of safety, efficacy, and utilization.

Real-world data, on the other hand, encompasses the raw data collected from various sources such as electronic health records (EHRs), insurance claims databases, patient registries, and observational studies. RWD serves as the foundation for generating RWE, but it becomes clinically relevant only after rigorous analysis, leading to actionable insights.

In contrast to RWE, *randomized controlled trials* are the gold standard for establishing the efficacy and safety of new pharmaceutical products. RCTs are conducted in highly controlled environments with strict inclusion and exclusion criteria, designed to minimize variability and bias. While RCTs provide critical evidence for regulatory approval, they may not fully reflect how a product performs in real-world settings where patient populations are more heterogeneous, and adherence to treatment may differ.

Key Differences Between RWE, RWD, and RCTs

- *RWE* is generated from the analysis of *RWD* to offer insights into treatment effectiveness in real-world settings.
- *RCTs* provide evidence of efficacy and safety under controlled conditions, but RWE complements RCTs by demonstrating how treatments perform in everyday practice.
- *RWD* is the raw, unprocessed data, while *RWE* is the interpretation of that data to answer clinical and business questions.

HEOR and its Relationship with RWE

HEOR is a field that focuses on the economic aspects of healthcare interventions, assessing the value of new drugs or treatments in terms of their cost-effectiveness, quality of life improvements, and health outcomes. HEOR studies help pharmaceutical companies, payers, and policymakers determine the best allocation of healthcare resources.

HEOR relies heavily on economic modeling and outcomes research, which may include data from both RCTs and real-world settings. HEOR is typically concerned with questions such as what is the cost per quality-adjusted life year (QALY) gained from using a specific therapy? How does the new treatment compare to existing therapies in terms of both clinical outcomes and economic value?

Difference Between RWE and HEOR

- *RWE* focuses on generating clinical evidence from RWD sources to assess the effectiveness and safety of therapies.
- *HEOR* is concerned with the economic and patient outcome aspects of healthcare interventions, incorporating data from both RWE and clinical trials.
- *Overlap*: While RWE provides the clinical insights, HEOR applies these insights to assess economic outcomes and inform decision-making for payers and policymakers.

Key Methodologies for Collecting and Analyzing RWE

The collection and analysis of RWE require robust methodologies to ensure the data is valid, reliable, and actionable. RWE studies often use observational designs, and the data can be sourced from a variety of real-world settings.
Here are the primary methods and data sources used in RWE generation:

Patient Registries

Patient registries are organized systems that collect data on patients with specific diseases or undergoing specific treatments. These registries provide valuable longitudinal data, allowing for the assessment of treatment outcomes over extended periods of time. Registries are particularly useful for studying rare diseases, where RCTs may be difficult to conduct due to the small patient population.

Example: The European Union's DARWIN (Data Analysis and Real-World Interrogation Network) initiative by the European Medicines Agency (EMA) is a notable example. DARWIN aims to use patient registries and other data sources to generate robust RWE on the performance of medicines across the EU. This initiative allows regulators, payers, and healthcare providers to access real-time data on the effectiveness, safety, and utilization of therapies, helping to inform regulatory and reimbursement decisions.

Electronic Health Records

EHRs are digitized medical records collected by healthcare providers during routine clinical care. EHRs are a rich source of RWD, capturing patient demographics, diagnosis codes, treatment information, and clinical outcomes. EHRs reflect routine clinical practice and, as such, provide a real-world view of how therapies are used and how patients respond to treatments.

Advantages
- Captures data on large and diverse patient populations.
- Reflects real-time clinical practices and patient outcomes.

Challenges
- Data quality can vary, and missing information can limit its usefulness.
- Integrating EHRs from different systems can be complex due to a lack of standardization.

Claims and Billing Data

- Healthcare claims and billing data are generated when healthcare providers submit claims to insurance companies for reimbursement. This data provides information on healthcare utilization, including hospital admissions, procedures performed, prescription drug use, and overall healthcare costs.

Advantages
- Comprehensive data on healthcare utilization and associated costs.
- Useful for health economics research and evaluating healthcare resource utilization.

Challenges
- Limited clinical detail as claims data primarily focuses on billing codes rather than patient outcomes.

Patient-Reported Outcomes (PROs)

PROs capture the patient's perspective on their health status, quality of life, and treatment satisfaction. These outcomes are particularly valuable for understanding how treatments impact patients' daily lives and overall well-being. PROs are often collected through surveys or mobile health apps.

Advantages
- Provides direct insights from patients on their experiences with treatments.
- Useful for evaluating outcomes that are not always captured in clinical trials, such as quality of life improvements.

Challenges
- PROs can be subjective and influenced by patient expectations or recall bias.

Observational Studies and Pragmatic Trials

Observational studies track patients in real-world settings without the intervention and control inherent in RCTs. These studies can be prospective or retrospective and provide data on how treatments perform across broader patient populations.

Pragmatic trials, a form of RWE study, incorporate elements of both RCTs and observational studies by embedding research into routine clinical practice.

Advantages

- Observational studies provide real-world insights into the effectiveness and safety of treatments.
- Pragmatic trials can generate high-quality evidence with the flexibility to incorporate real-world variables.

Challenges

- Observational studies are prone to bias due to lack of randomization.
- Data interpretation can be complex due to confounding factors.

The Pivotal Role of Medical Affairs in RWE

Medical Affairs teams play a critical role in the generation, interpretation, and application of RWE. Their involvement encompasses all aspects of RWE from the design of RWE studies to the communication of findings with both internal and external stakeholders.

How Medical Affairs adds value at each stage of the RWE lifecycle:

Study Design and Data Collection

Medical Affairs is often responsible for designing RWE studies to address specific scientific or business objectives. This can involve collaborating with other departments (e.g., Market Access, HEOR) to ensure that the study is aligned with regulatory and payer requirements. Medical Affairs helps ensure that RWE studies are methodologically sound and that they address meaningful clinical questions.

Example: When a pharmaceutical company aims to expand the indication of an existing therapy, Medical Affairs may design an observational study to collect data on the product's real-world use in a new patient population. This data can be pivotal in supporting regulatory submissions and gaining approval for label expansion.

Data Interpretation and Analysis

Medical Affairs professionals bring their clinical and scientific expertise to the interpretation of RWE. This is especially important because RWD can be messy and inconsistent. Medical Affairs helps ensure that the data is analyzed rigorously, identifying trends, gaps, or safety signals that may influence business strategies.

Example: After analyzing data from an EHR-based study, Medical Affairs identifies a trend where a particular subgroup of patients responds better to a treatment than the broader population. This insight can inform future clinical trial designs, marketing strategies, and healthcare provider education.

Regulatory and Payer Interactions

RWE plays an increasingly important role in interactions with regulatory authorities and payers. Medical Affairs helps translate RWE into a narrative that can support regulatory filings, health technology assessments (HTAs), and reimbursement discussions. RWE can demonstrate long-term safety and effectiveness, especially in cases where RCTs may not fully capture the broader population.

Example: The DARWIN initiative by the EMA utilizes RWE to support post-market surveillance, and Medical Affairs teams are often involved in ensuring that the data from such initiatives is communicated effectively to regulators and payers.

Communication with Healthcare Professionals and Patients

Medical Affairs is responsible for communicating the findings of RWE studies to healthcare professionals (HCPs), patients, and other stakeholders. This may involve presenting RWE at medical conferences, publishing findings in peer-reviewed journals, or creating educational materials for healthcare providers.

Example: A company launches a new educational campaign for oncologists based on RWE showing the comparative effectiveness of its oncology drug versus existing treatments. Medical Affairs plays a key role in ensuring that the data is communicated clearly and that it is scientifically accurate.

Supporting Market Access and Commercial Strategy

In collaboration with Market Access teams, Medical Affairs ensures that RWE is used to support reimbursement and pricing strategies. RWE is often critical for demonstrating value to payers, particularly when pricing decisions hinge on the real-world cost-effectiveness of a therapy.

Example: A pharmaceutical company uses RWE to demonstrate that its new biologic reduces hospital readmissions and long-term healthcare costs compared to standard care. This data is pivotal in securing favorable pricing and reimbursement decisions from payers.

Implications of RWE for Pharmaceutical Business

The growing importance of RWE in the pharmaceutical industry has several key implications for business strategy, product development, and regulatory interactions:

- *Accelerating Regulatory Approvals*: Regulatory agencies such as the FDA and EMA are increasingly relying on RWE to supplement RCT data, particularly for post-approval studies or label expansions. This trend allows pharmaceutical companies to expedite regulatory approvals by demonstrating real-world effectiveness and safety across diverse populations.
- *Enhancing Market Access and Reimbursement*: RWE helps pharmaceutical companies present a comprehensive value proposition to payers, demonstrating not only clinical efficacy but also long-term cost-effectiveness. By leveraging RWE, companies can improve market access, secure favorable pricing, and ensure broader reimbursement coverage.
- *Driving Product Innovation*: RWE can provide insights that inform product development and lifecycle management. For example, RWE might identify unmet clinical needs or new therapeutic applications, which can drive the innovation pipeline. Furthermore, data from real-world studies can be used to refine dosing regimens, improve patient selection criteria, or inform patient support programs.
- *Mitigating Risk and Improving Pharmacovigilance*: RWE plays a crucial role in post-market surveillance, helping pharmaceutical companies monitor product safety and efficacy over the long term. By identifying adverse events or trends in specific patient populations, companies can mitigate risks and take proactive measures to enhance patient safety and product stewardship.

RWE is becoming a cornerstone of pharmaceutical decision-making, offering invaluable insights into how therapies perform in real-world clinical practice. The role of Medical Affairs is pivotal in ensuring that RWE is collected, interpreted, and applied in a way that drives business success while maintaining scientific integrity. By fostering collaboration across departments, supporting regulatory and payer interactions, and communicating insights to HCPs, Medical Affairs ensures that RWE contributes to a company's long-term growth and success in the marketplace. As initiatives like DARWIN demonstrate, the future of pharmaceutical development will increasingly rely on the effective use of RWE, and Medical Affairs will remain at the forefront of this evolution.

Special Focus on Legal and Ethical Considerations in the Use of RWE

The use of RWD to generate RWE poses several legal and ethical challenges that must be carefully navigated. Pharmaceutical companies, guided by their Medical Affairs teams, must ensure that RWE studies comply with data privacy laws, patient consent regulations, and ethical standards for human research. Failure to do so can result in legal penalties, loss of stakeholder trust, and even harm to patients.

Data Privacy and Security

Data privacy is a critical concern when utilizing RWD, especially in regions with stringent regulations such as the European Union (EU) and the United States. In the EU, the General Data Protection Regulation (GDPR) provides strict guidelines on how personal data can be collected, stored, and used. Similarly, in the United States, the Health Insurance Portability and Accountability Act (HIPAA) governs the handling of patient data.

Role of Medical Affairs in Data Privacy

Medical Affairs plays a central role in ensuring that all RWE studies comply with these data privacy laws. For example, when RWD is collected from patient registries or EHRs, Medical Affairs must ensure that the data is anonymized or de-identified to protect patient confidentiality. Additionally, Medical Affairs teams work closely with legal and compliance departments to establish data-sharing agreements that comply with regulatory frameworks.

Ethical Use of Patient Data

Beyond legal obligations, there is an ethical imperative to use patient data responsibly. In many cases, patients may not be aware that their health data is being used for secondary purposes like RWE generation. Medical Affairs teams must ensure that informed consent is obtained where necessary and that the use of data is transparent, with patients and stakeholders understanding how the data will be applied.

Informed Consent and Patient Autonomy

The principle of informed consent is a cornerstone of ethical medical research, and it applies equally to RWE studies. While consent may not always be required for the use of anonymized data in observational studies, certain jurisdictions mandate explicit consent for specific uses of health information.

Example: In observational studies that collect PROs, patients are directly involved in providing information about their health and quality of life. In such cases, Medical Affairs ensures that patients are fully informed about how their data will be used, what risks are involved, and how their privacy will be protected.

Special Focus on Leveraging Technology in RWE Collection and Analysis

The growing availability of health data from EHRs, wearables, and other digital sources is transforming how pharmaceutical companies collect and analyze RWE. New technologies, such as artificial intelligence (AI) and machine learning (ML), are enabling Medical Affairs to extract insights from vast datasets, providing deeper understanding and more predictive models of real-world treatment outcomes.

Artificial Intelligence and Machine Learning in RWE

AI and ML have revolutionized how pharmaceutical companies handle large datasets, improving the efficiency and accuracy of RWE generation. These technologies allow Medical Affairs teams to process and analyze RWD in ways that were previously impossible using traditional methods. AI-driven algorithms can identify patterns and trends in patient outcomes, helping to predict which therapies are most effective for specific populations.

Example: AI tools can analyze EHRs to track disease progression and patient responses to therapy over time, uncovering real-world treatment pathways that differ from those observed in clinical trials. This information can be used to optimize future clinical trial designs or to identify new therapeutic targets.

Benefits of AI in RWE

- Enhanced ability to analyze unstructured data (e.g., doctor's notes, imaging reports).
- Identification of subpopulations that benefit most from specific treatments.
- Early detection of safety signals or adverse events.

Big Data Analytics and Integration of Diverse Data Sources

RWE generation often involves integrating data from multiple sources, including EHRs, patient registries, claims data, and PROs. Advanced data analytics tools enable Medical Affairs teams to combine and analyze these diverse datasets, producing a more comprehensive view of treatment outcomes.

Example: By merging data from patient registries and claims databases, pharmaceutical companies can gain insights into both the clinical outcomes and economic impact of their treatments. This integration can inform HEOR as well as market access strategies.

Challenges

- Data heterogeneity: Differences in coding practices, missing data, or inconsistent follow-up periods can complicate data integration.
- Interoperability: Ensuring that different data systems can communicate effectively remains a significant technical hurdle.

Special Focus on Challenges and Limitations of RWE

While RWE provides a wealth of opportunities for understanding treatment effectiveness in real-world settings, it also has inherent challenges and limitations. These issues can affect the validity and reliability of RWE studies and must be carefully managed by Medical Affairs teams to ensure that the findings are robust and actionable.

Data Quality and Completeness

RWD, particularly from sources such as EHRs and claims data, can often be incomplete or inconsistent. Unlike RCTs, which follow strict protocols for data collection, RWE relies on routine clinical practice, where data capture may vary widely. Missing data, irregular follow-up periods, and inconsistent coding practices can all impact the quality of RWE.

Example: A study using EHR data might suffer from gaps in patient records if patients switch healthcare providers, leading to incomplete treatment histories. Medical Affairs teams must develop strategies for dealing with such gaps, such as using statistical methods to impute missing values or working with multiple data sources to validate the findings.

Mitigating Strategies

- *Data Cleaning*: Medical Affairs teams often work closely with data scientists to standardize and clean RWD before analysis.
- *Sensitivity Analysis*: Performing sensitivity analyses to assess the robustness of the results when key variables are missing or inconsistent.

Confounding Factors and Bias

Since RWE is collected from observational studies, there is a risk of bias and confounding factors that can influence the outcomes. For example, patients receiving a particular treatment may differ from those receiving another therapy due to differences in health status, socioeconomic factors, or healthcare access. Unlike RCTs, where randomization controls for these variables, RWE studies must account for these confounding factors through careful study design and analysis.

Mitigating Strategies

- *Propensity Score Matching*: This statistical technique is often used to create comparable groups of patients based on baseline characteristics, thereby reducing bias.
- Multivariate Analysis: Medical Affairs teams may use multivariate regression models to control for potential confounders and better isolate the treatment effect.

Interpretation and Generalizability

The interpretation of RWE can be challenging, particularly when attempting to generalize the findings to broader populations. While RWE reflects real-world practice, the data may not always be representative of the entire population due to selection biases or variations in healthcare access.

Example: A RWE study conducted using data from a specific geographic region or healthcare system may not be applicable to other regions with different healthcare practices or patient demographics. Medical Affairs must carefully consider the generalizability of RWE findings when making recommendations or applying the results to broader clinical and business strategies.

RWE is transforming the pharmaceutical industry by providing insights into how therapies perform in everyday clinical practice. However, leveraging RWE effectively requires addressing several challenges related to data quality, bias, and legal and ethical considerations. Medical Affairs plays a pivotal role in ensuring that RWE is collected, analyzed, and applied responsibly, guiding pharmaceutical companies toward more informed decisions. By harnessing new technologies and ensuring compliance with regulatory frameworks, Medical Affairs teams can maximize the value of RWE and help shape the future of drug development and patient care.

Case Study: The DARWIN Initiative to Leverage RWE for Regulatory Decision-Making (https://www.ema.europa.eu/en/about-us/how-we-work/data-regulation-big-data-other-sources/realworld-evidence/data-analysis-real-world-interrogation-network-darwin-eu)

The *Data Analysis and Real-World Interrogation Network (DARWIN)* is an ambitious initiative spearheaded by the *EMA* with the primary objective of harnessing the power of RWE to inform regulatory decision-making. DARWIN aims to integrate RWD from various healthcare sources across Europe, creating a network that provides regulators, healthcare providers, and pharmaceutical companies with access to high-quality, real-time data on the utilization, safety, and effectiveness of medicines in everyday clinical practice. DARWIN represents a transformative approach to post-market surveillance and regulatory oversight, reinforcing the EMA's commitment to improving patient outcomes and enhancing the safety and efficacy of medicinal products.

Purpose of the DARWIN Initiative

The primary purpose of DARWIN is to support regulatory decisions throughout the lifecycle of medicines by leveraging RWE. The initiative aims to:

- *Enhance Post-market Surveillance*: By using RWE, DARWIN improves the ability to monitor the safety and effectiveness of medicines once they are on the market, offering insights into how drugs perform across diverse and heterogeneous patient populations.
- *Inform Regulatory Decisions*: DARWIN helps regulators make more informed decisions regarding drug approvals, label expansions, and safety assessments. It provides a robust framework for real-time data analysis, ensuring that decisions are based on the most current and comprehensive evidence available.
- *Support Health Technology Assessments (HTAs)*: DARWIN enables health technology assessment bodies to access RWD that can be used to evaluate the cost-effectiveness and real-world impact of new and existing therapies, facilitating better-informed decisions about reimbursement and pricing.
- *Drive Innovation in Regulatory Science*: The initiative encourages innovation by incorporating new methodologies for data collection, sharing, and analysis. It aims to standardize the use of RWE across Europe, enabling a more coordinated and efficient approach to regulatory science.

Structure of DARWIN

The DARWIN initiative is built around a networked infrastructure that connects various data sources across Europe. It aims to create an integrated system that pools data from hospitals, EHRs, registries, and other healthcare settings, enabling large-scale analysis of real-world outcomes. The structure of DARWIN involves multiple layers of collaboration, ensuring that the network is comprehensive, secure, and efficient in its operations.

- *Centralized Data Hub*: At the core of DARWIN is a centralized data hub that facilitates access to RWD from multiple sources. This hub acts as the central repository where RWD from different healthcare systems across Europe is processed, standardized, and made accessible to authorized stakeholders.
- *Data Sources*: DARWIN's data sources are varied and include:

  1. *EHRs*: Capturing real-time clinical data from patient interactions with healthcare systems.
  2. *Patient Registries*: Providing structured, longitudinal data on patients with specific diseases or conditions.
  3. *Health Insurance Claims Databases*: Offering data on healthcare utilization, treatments, and associated costs.
  4. *Pharmacovigilance Databases*: Collecting data on adverse events and other safety concerns related to pharmaceutical products.

- *Data Standardization and Quality Control*: To ensure the consistency and reliability of the data used for RWE, DARWIN incorporates rigorous data standardization protocols. Data from various healthcare settings are harmonized to create a standardized format that allows for seamless analysis across different countries

and health systems. A strong focus is placed on data quality, with systems in place for regular auditing and validation of the data.
- *Secure Data Access*: Security and data privacy are paramount in DARWIN, given the sensitive nature of healthcare data. The initiative incorporates advanced security measures, including data anonymization and encryption, ensuring that patient confidentiality is maintained. Only authorized stakeholders, such as regulatory agencies, pharmaceutical companies, and academic researchers, are granted access to the data, and strict data governance rules are enforced.

Key Stakeholders in DARWIN

The success of DARWIN hinges on the collaboration of multiple stakeholders who contribute to, manage, and benefit from the initiative. These stakeholders include regulatory bodies, pharmaceutical companies, HCPs, and patients, each playing a unique role in the network.

- *EMA*: As the leading regulatory authority behind DARWIN, the EMA is responsible for overseeing the initiative, ensuring that the data collected is utilized to inform regulatory decisions. The EMA leads the governance structure of DARWIN, ensuring that the data analysis aligns with regulatory science needs, such as post-approval surveillance and safety assessments.
- *National Competent Authorities (NCAs)*: NCAs from various European countries participate in DARWIN by providing access to national healthcare data, contributing to a more comprehensive view of real-world outcomes across the EU. These authorities also help enforce data privacy regulations, ensuring compliance with national and EU-wide laws such as the General Data Protection Regulation (GDPR).
- *Pharmaceutical Industry*: Pharmaceutical companies are key stakeholders in DARWIN, as they stand to benefit significantly from access to RWE generated by the network. Companies can use the data to support regulatory submissions, post-market surveillance, and health technology assessments. RWE generated from DARWIN can also guide strategic business decisions, such as market expansion or label extensions.
- *Healthcare Providers and Hospitals*: Hospitals and HCPs contribute RWD to the DARWIN network through their EHR systems and patient registries. These providers benefit from insights into how medicines are performing in clinical practice and can use the data to improve patient outcomes.
- *Patients and Patient Advocacy Groups*: Patients play a central role in the success of DARWIN by contributing their healthcare data, either directly (through registries) or indirectly (through EHRs). Patient advocacy groups work closely with the EMA and other stakeholders to ensure that patient rights are respected, particularly regarding data privacy and informed consent. DARWIN aims to provide patients with safer, more effective treatments by using RWE to enhance regulatory decision-making.
- *Health Technology Assessment (HTA) Bodies*: HTA agencies use data from DARWIN to assess the cost-effectiveness and value of new treatments. Access to RWE enables these bodies to make more informed decisions about the reimbursement and pricing of therapies, ensuring that healthcare systems allocate resources effectively.

Applications and Benefits of DARWIN

The DARWIN initiative offers numerous applications that benefit a wide range of stakeholders, enhancing the overall landscape of regulatory science and healthcare innovation.

- *Post-Market Surveillance*: One of the primary uses of DARWIN is to improve post-market surveillance of medicines. By accessing RWD in real-time, regulators can monitor the safety and effectiveness of therapies long after they have been approved. This continuous surveillance allows for the identification of adverse events that may not have been detected in clinical trials, improving patient safety.
- *Supporting Regulatory Decisions*: DARWIN provides regulators with comprehensive data on how medicines perform across different patient populations and healthcare settings. This data is invaluable for making informed decisions about new approvals, label expansions, or safety-related updates. RWE generated from DARWIN can serve as complementary evidence to RCTs, offering a broader view of a product's real-world impact.
- *Health Technology Assessments*: By providing detailed RWD on the effectiveness and cost of treatments, DARWIN supports HTAs that inform reimbursement and pricing decisions. HTA bodies can use this data to evaluate the long-term value of therapies, taking into account real-world patient outcomes and healthcare costs.
- *Public Health Decision-Making*: DARWIN's comprehensive database can also be used to inform public health strategies. Governments and public health authorities can leverage the network's data to identify trends in disease prevalence, treatment effectiveness, and healthcare utilization, guiding policy decisions and resource allocation.
- *Facilitating Research and Innovation*: DARWIN's rich data repository provides an invaluable resource for pharmaceutical companies, academic researchers, and HCPs to explore new avenues of research. Insights gained from the network can guide drug development, inform clinical trial designs, and promote innovation in treatment approaches.

Challenges and Future Directions

While DARWIN offers a promising framework for the use of RWE in regulatory decision-making, there are also challenges that must be addressed. Data standardization across multiple healthcare systems remains complex, and ensuring interoperability between different data sources can be difficult. Moreover, privacy concerns continue to be a key consideration, with ongoing efforts to ensure compliance with regulations like the GDPR.

Looking ahead, DARWIN aims to expand its network, incorporating more data sources and countries to create an even more robust repository of RWE. The initiative is also expected to play an increasingly important role in the EMA's digital strategy, as the agency explores new ways to integrate advanced analytics, AI, and ML into regulatory processes.

The DARWIN initiative represents a groundbreaking step forward in the use of RWE to inform regulatory decisions and improve patient outcomes. By integrating RWD from healthcare systems across Europe, DARWIN provides regulators, HCPs, and pharmaceutical companies with the insights needed to make more informed decisions about the safety, effectiveness, and cost of medicines. Medical Affairs teams within pharmaceutical companies are key beneficiaries of DARWIN, as the initiative offers access to high-quality data that can be used to support regulatory submissions, HTAs, and post-market surveillance efforts. As DARWIN continues to evolve, it is expected to play an increasingly central role in shaping the future of regulatory science and healthcare innovation in Europe.

Chapter 12
Ethics and Compliance: The Leading Role of Medical Affairs

Key Highlights
- Medical Affairs safeguards ethical integrity in scientific communication and stakeholder interactions.
- Compliance frameworks (Sunshine Act, EFPIA, GDPR) enforce transparency in financial disclosures and patient data protection.
- Ethical marketing practices ensure patient welfare is prioritized over commercial interests.
- AI and digital health introduce new ethical challenges, including data security and bias mitigation.
- Ethical decision-making in clinical trials and regulatory interactions remains crucial for industry credibility.

Ethics and compliance form the bedrock of pharmaceutical operations, ensuring that every facet of drug development, commercialization, and interaction with healthcare professionals (HCPs) adheres to strict regulatory frameworks and high moral standards. The evolving regulatory landscape, with growing demands for transparency and accountability, has placed ethics and compliance at the forefront of pharmaceutical practices. Among the many functions within a pharmaceutical company, Medical Affairs has emerged as a critical leader in maintaining ethical standards and ensuring compliance across various activities, from scientific communication to patient safety and promotional practices.

Definition of Ethics and Its Importance in Healthcare

Ethics, broadly defined, refers to the moral principles that govern behavior. In the context of healthcare, ethics guide decisions and actions to prioritize patient well-being, ensure fairness, and uphold the trust placed in HCPs and the pharmaceutical

A. Krendyukov, *Medical Affairs' Fundamentals: The Next Generation*,
https://doi.org/10.1007/978-3-031-92588-7_12

industry. Ethical principles are fundamental to every aspect of drug development, clinical trials, commercialization, and post-market surveillance.

The Hippocratic Oath and its Legacy

The Hippocratic Oath, one of the earliest expressions of medical ethics, embodies the core principle of "do no harm" and continues to influence the ethical frameworks that guide HCPs today. While this ancient oath focuses primarily on the duties of individual practitioners, its central tenets have expanded to shape the moral obligations of modern pharmaceutical companies.

In the pharmaceutical sector, ethics emphasize patient safety, transparency in scientific communication, and the responsible development and marketing of medicines. These ethical obligations apply to all stages of the drug lifecycle, from clinical trials to the eventual sale of a product.

Current Regulations in Drug Development and Commercialization

In the modern era, healthcare ethics have become closely intertwined with regulatory frameworks designed to ensure the safety, efficacy, and ethical marketing of pharmaceutical products. The pharmaceutical industry operates under stringent regulations enforced by agencies such as the U.S. Food and Drug Administration (FDA) and the European Medicines Agency (EMA). These regulations govern every aspect of drug development, from preclinical research and clinical trials to labeling, post-market surveillance, and promotional activities.

In addition to regulatory oversight, ethical considerations extend to the interactions between pharmaceutical companies and HCPs. These interactions must be transparent and free from conflicts of interest to maintain public trust and ensure that patients receive treatments based on clinical need rather than financial incentives.

The Evolving Role of Compliance Over the Last Decades

Compliance refers to the adherence to laws, regulations, and industry codes of conduct. In the pharmaceutical sector, compliance ensures that companies act in accordance with ethical standards and legal requirements throughout the drug development and commercialization process. Over the last few decades, the role of compliance has evolved significantly, driven by growing scrutiny of pharmaceutical practices and an increasing emphasis on transparency and patient protection.

The Sunshine Act (USA) and European Regulations

One of the most significant regulatory developments in recent years is the Sunshine Act, a U.S. law enacted in 2010 under the Patient Protection and Affordable Care Act. The Sunshine Act requires pharmaceutical companies to publicly disclose any payments or transfers of value made to HCPs or organizations. The purpose of the Sunshine Act is to increase transparency in the financial relationships between pharmaceutical companies and HCPs, ensuring that any interactions are free from undue influence and do not compromise clinical decision-making.

In Europe, transparency regulations are similarly stringent, with the EMA playing a leading role in enforcing ethical standards. In addition to requiring robust clinical trial transparency, the EMA has implemented regulations that govern the promotion and marketing of medicines. These regulations ensure that pharmaceutical companies provide accurate, balanced, and evidence-based information about their products to both HCPs and the public.

Impact of Evolving Compliance Regulations on the Pharmaceutical Industry

The introduction of laws such as the Sunshine Act and similar transparency regulations in the European Union has had a profound impact on the pharmaceutical industry. Companies are now required to establish comprehensive compliance programs to monitor and report interactions with HCPs, manage conflicts of interest, and ensure that marketing practices adhere to ethical guidelines. These regulatory developments have driven the need for robust internal controls and a culture of ethical behavior across all functions, with Medical Affairs playing a leading role in ensuring compliance.

Current Trends in Stringent Regulations, Transparency, and Code of Conduct

The pharmaceutical industry faces increasing demands for transparency and ethical conduct, driven by regulatory bodies, patient advocacy groups, and the public. As part of these demands, pharmaceutical companies must comply with anti-bribery, anti-corruption, and conflict of interest laws to prevent undue influence on HCPs and ensure that clinical decisions are based on patient needs rather than financial incentives.

Transparency in Financial Relationships with Healthcare Professionals

Financial relationships between pharmaceutical companies and HCPs have long been a focus of regulatory scrutiny. Transparency laws, such as the Sunshine Act, require full disclosure of any payments, gifts, or other transfers of value from pharmaceutical companies to HCPs. These regulations aim to prevent unethical practices such as bribery or inducements, which could lead to biased prescribing behavior or the promotion of products for financial gain rather than clinical merit.

Codes of Conduct and Anti-Corruption Laws

To address concerns about corruption and unethical practices, many countries have implemented anti-bribery and anti-corruption laws that govern the interactions between pharmaceutical companies and HCPs. These laws prohibit companies from offering financial incentives to HCPs in exchange for prescribing their products. Industry associations, such as the Pharmaceutical Research and Manufacturers of America (PhRMA) and the International Federation of Pharmaceutical Manufacturers & Associations (IFPMA), have developed strict codes of conduct to provide guidance on ethical interactions between companies and HCPs.

Ethical Considerations in Medical Affairs

Ethical considerations are central to the function of Medical Affairs in the pharmaceutical industry. Upholding high ethical standards is imperative not only for compliance with regulatory requirements but also for maintaining public trust and ensuring patient safety. Medical Affairs teams are responsible for navigating complex ethical dilemmas and ensuring that all scientific communications, interactions with HCPs, and patient-related activities are conducted with transparency and integrity.

Transparency in Scientific Communication

- *Disclosure of Conflicts of Interest*: Conflicts of interest, whether real or perceived, can significantly undermine the credibility of Medical Affairs. A key ethical obligation is the transparent disclosure of any conflicts of interest. This includes financial relationships, affiliations, or personal interests that could be seen as influencing the objectivity of information or decisions made by Medical Affairs professionals.

Example: When a Medical Affairs professional publishes a scientific paper or presents data at a conference, it is crucial to disclose any relationships with the pharmaceutical company, other research entities, or financial stakeholders. This transparency ensures that the audience can evaluate the information presented with full knowledge of any potential biases.

- *Maintaining Integrity in the Dissemination of Scientific Information*: Integrity in the dissemination of scientific information is another cornerstone of ethical practice in Medical Affairs. This involves presenting data in an unbiased, accurate, and balanced manner. Selective reporting of results, overstatement of benefits, or understatement of risks can lead to misinformation and potentially harm patients.

Role of Medical Affairs: Medical Affairs teams are responsible for ensuring that all scientific communications—whether in educational materials, marketing content, or responses to inquiries—adhere to evidence-based medicine principles and maintain the highest standards of transparency.

Patient Privacy and Data Security

- *Safeguarding Patient Information in Medical Affairs Activities:* The increasing collection and analysis of patient data present ethical challenges, particularly with respect to privacy and data security. Medical Affairs must ensure compliance with laws such as the General Data Protection Regulation (GDPR) in the European Union and the Health Insurance Portability and Accountability Act (HIPAA) in the United States. These regulations require companies to protect sensitive patient information, ensure that data is used only for authorized purposes, and obtain informed consent for data collection and usage.

Example: When conducting post-marketing surveillance or real-world evidence (RWE) studies, Medical Affairs must ensure that patient data is anonymized or de-identified to protect patient confidentiality.

- *Adhering to Ethical Standards in Data Collection and Analysis*: Adhering to ethical standards in data collection and analysis is essential for maintaining the integrity of Medical Affairs activities. This includes ensuring that data is collected with respect for patient rights and that the methodologies used are scientifically sound. Transparency about the limitations of data and avoiding overinterpretation of results is also critical in upholding ethical standards.

Ethical Marketing and Promotional Practices

- *Ensuring Ethical Interactions with Healthcare Professionals*: Medical Affairs plays a crucial role in ensuring that all interactions between the pharmaceutical company and HCPs are conducted ethically. This includes adhering to industry codes of conduct, such as the PhRMA Code or the IFPMA Code of Practice, which provide guidance on the appropriate provision of educational and promotional materials.
- *Balancing Commercial Interests with Patient Welfare*: One of the greatest ethical challenges for Medical Affairs is balancing commercial interests with patient welfare. While pharmaceutical companies are for-profit entities, the primary mission of Medical Affairs is to ensure that patients receive the best possible care. Medical Affairs must advocate for decisions that prioritize patient safety, even when those decisions conflict with short-term commercial goals.

Ethical Considerations in Clinical Trials

- *Upholding Ethical Standards in Clinical Research*: Clinical trials are a cornerstone of pharmaceutical development, and Medical Affairs plays a vital role in ensuring that these trials adhere to ethical standards. This includes safeguarding patient safety, obtaining informed consent, and respecting patient autonomy throughout the research process.

Example: Medical Affairs teams ensure that clinical trials are designed to answer meaningful scientific questions and that the benefits of research justify any potential risks to participants. They are also responsible for ensuring transparency in the publication of clinical trial results.

- *Addressing Ethical Dilemmas in Clinical Decision-Making*: Medical Affairs professionals often face ethical dilemmas when interpreting ambiguous data or balancing competing interests. Addressing these dilemmas requires a careful, evidence-based approach and may involve consulting ethics committees or patient advocacy groups to ensure that decisions prioritize patient welfare.

Ethics and compliance are integral to the functioning of pharmaceutical companies, and Medical Affairs is at the forefront of ensuring that all activities align with these principles. From transparency in scientific communication to safeguarding patient privacy, Medical Affairs serves as the ethical compass that guides the pharmaceutical industry through complex challenges. As regulatory frameworks evolve and the public demands greater transparency and accountability, the role of Medical Affairs in ensuring both ethics and compliance will continue to grow in importance. By upholding high ethical standards, Medical Affairs can help the pharmaceutical industry maintain public trust, improve patient outcomes, and navigate the ever-changing regulatory landscape.

Special Focus on Emerging Ethical Challenges in the Age of Digital Health and Personalized Medicine: The Role of Medical Affairs

The pharmaceutical industry is entering an era defined by rapid technological advancement and unprecedented access to patient data. Digital health technologies, artificial intelligence (AI), machine learning (ML), and personalized medicine are revolutionizing how healthcare is delivered and how pharmaceutical companies operate. However, with these advancements come new ethical challenges that Medical Affairs must address. The role of Medical Affairs is not only to ensure compliance with evolving regulations but also to act as the ethical compass for the pharmaceutical industry.

Emerging Ethical Challenges in the Age of Digital Health and Personalized Medicine

The rise of digital health technologies and personalized medicine has introduced a host of ethical challenges for the pharmaceutical industry. These innovations offer significant benefits, from improving patient outcomes to streamlining drug development, but they also present complex questions around patient privacy, consent, data security, and the potential for bias in healthcare delivery. Medical Affairs plays a central role in navigating these challenges, ensuring that the implementation of new technologies is ethical and that patient welfare remains paramount.

AI and ML have the potential to revolutionize healthcare decision-making, from drug discovery to patient treatment plans. However, the use of algorithms in healthcare raises concerns about transparency, bias, and accountability.

- *Transparency and Accountability in AI Systems*: One of the primary ethical concerns surrounding AI in healthcare is the lack of transparency in how algorithms make decisions. AI systems are often seen as "black boxes," meaning that their decision-making processes are not easily understood by humans. This opacity can lead to mistrust, especially when decisions about patient care are involved. Medical Affairs teams must ensure that AI-driven tools used in pharmaceutical applications are transparent, with clear explanations of how data is processed and how decisions are reached.

Example: A pharmaceutical company develops an AI-driven diagnostic tool that helps physicians choose the best treatment for cancer patients. Medical Affairs must ensure that the algorithm is tested rigorously for accuracy and explain its decision-making process to HCPs, ensuring they understand the rationale behind its recommendations.

- *Bias in AI and Its Impact on Healthcare*: AI algorithms are only as good as the data they are trained on, which means they are susceptible to bias. *Bias in AI can* lead to unequal treatment outcomes, particularly for underrepresented groups. Medical Affairs has an ethical responsibility to ensure that AI tools used in healthcare are free from bias and do not perpetuate health disparities. This requires careful validation of AI models, ensuring that they are trained on diverse datasets that represent all patient populations.

  Example: An AI tool developed for predicting drug efficacy may underperform in minority populations if the training data lacks representation from these groups. Medical Affairs must work to ensure that diverse patient populations are included in clinical trials and AI training datasets to mitigate this risk.

Personalized medicine tailors treatments to individual patients based on their genetic profiles, improving the efficacy and safety of therapies. However, the use of genetic data in drug development and patient care raises ethical concerns about privacy, consent, and data ownership.

- *Privacy and Consent in the Use of Genetic Data*: Genetic data is highly sensitive, and its use in personalized medicine requires careful attention to privacy and informed consent. Patients must understand how their genetic information will be used, who will have access to it, and how it will be protected. Medical Affairs plays a critical role in ensuring that informed consent processes are clear and that patients are fully aware of the potential risks and benefits of providing their genetic data for research and treatment purposes.

  Example: A pharmaceutical company developing a gene therapy for a rare genetic disorder collects genetic data from patients as part of a clinical trial. Medical Affairs must ensure that the informed consent forms are comprehensive, explaining how the data will be used, stored, and shared with third parties, and that patients are aware of their rights to withdraw consent.
- *Data Ownership and Control*: One of the key ethical questions in personalized medicine is who owns the genetic data. Patients may feel uncomfortable with the idea that pharmaceutical companies or other entities could profit from their genetic information without their explicit consent or benefit-sharing. Medical Affairs must navigate this sensitive issue by advocating for ethical data-sharing practices that respect patients' rights and ensure that they retain control over their genetic information.

Expanding the Role of Medical Affairs in Handling Ethical Dilemmas

Medical Affairs frequently encounters ethical dilemmas that require a careful balance between patient welfare, regulatory requirements, and commercial interests. This section explores how Medical Affairs professionals manage these competing priorities in real-world situations, with practical examples and case studies that demonstrate ethical decision-making in action.

Case Study: Handling Off-Label Promotion Requests
One of the most common ethical dilemmas in Medical Affairs arises when HCPs request information about the off-label use of a drug. While these requests are often legitimate, Medical Affairs must navigate strict regulations that prohibit promoting unapproved uses of pharmaceuticals. The challenge lies in providing objective, scientifically grounded information without violating compliance rules.

Scenario: Off-Label Use in Oncology

A physician contacts a pharmaceutical company's Medical Affairs team to inquire about using an approved oncology drug for an unapproved indication. While the physician has valid clinical reasons for considering this off-label use, Medical Affairs must ensure that its response is compliant with regulations.

Medical Affairs Response: The Medical Affairs team can provide the physician with a summary of relevant clinical data, including any peer-reviewed studies that mention the off-label use. However, they must clearly state that the drug is not approved for that specific indication and that the physician should consult regulatory guidelines before proceeding. This balanced approach allows Medical Affairs to provide valuable scientific information while maintaining regulatory compliance.

Balancing Commercial Interests and Patient Welfare

Medical Affairs professionals often face situations where commercial pressures—such as the desire to increase market share or accelerate product uptake—conflict with the ethical imperative to prioritize patient safety and well-being. In these cases, Medical Affairs must advocate for decisions that align with the best interests of patients, even when doing so may delay commercial goals.

Example: A company is preparing to launch a new biologic treatment, but early post-market surveillance data suggests that some patients are experiencing severe adverse reactions. Despite the commercial importance of the drug's launch, Medical Affairs recommends issuing an updated safety warning to HCPs and delaying further promotional activities until the issue is resolved.

Navigating Global Differences in Ethical Standards

Pharmaceutical companies operate in a global marketplace, and ethical standards and regulations can vary significantly between regions. Medical Affairs plays a crucial role in ensuring that the company complies with both global regulations and local ethical guidelines.

Ethical Standards in the United States and Europe

In the United States, pharmaceutical companies must adhere to regulations such as the Sunshine Act, which mandates transparency in financial interactions with HCPs. The Pharmaceutical Research and Manufacturers of America (PhRMA) Code provides additional guidelines on ethical marketing practices and interactions with HCPs.

In the European Union, the EMA oversees ethical standards for drug development and marketing, with a strong emphasis on transparency and patient safety. The European Federation of Pharmaceutical Industries and Associations (EFPIA) Code sets ethical standards for interactions between pharmaceutical companies and HCPs, similar to the PhRMA Code.

Global Differences: While the United States and Europe have well-established ethical frameworks, emerging markets may have less stringent regulations or inconsistent enforcement. Medical Affairs teams must take a proactive approach in these regions, ensuring that the company adheres to its global ethical standards even when local regulations are less comprehensive.

Measuring and Monitoring Ethical Compliance

To ensure that ethical standards are consistently upheld, pharmaceutical companies must implement systems for measuring and monitoring compliance. Medical Affairs plays a key role in tracking compliance metrics and providing measurable outcomes that demonstrate the effectiveness of ethics and compliance programs. This section outlines key performance indicators (KPIs) used to assess ethical compliance in pharmaceutical companies.

Key Performance Indicators (KPIs) for Ethical Compliance

- *Number of Conflict of Interest Disclosures*: One critical measure of ethical compliance is the number of conflict of interest disclosures. Medical Affairs tracks how often conflicts are disclosed and ensures that these disclosures are managed appropriately to prevent any undue influence on clinical decision-making or scientific communication.
- *Compliance Training Completion Rates*: Ethical behavior in pharmaceutical companies begins with a well-informed workforce. Tracking compliance training completion rates ensures that all employees, including Medical Affairs professionals, are adequately trained in the latest ethical standards and regulations.
- *Patient Privacy Incidents*: With the increasing collection and analysis of patient data, tracking the number of patient privacy incidents is a crucial metric. Medical Affairs teams must ensure that privacy breaches are minimized and that any incidents are addressed promptly and transparently.

The Future of Ethics and Compliance in Pharmaceutical Companies

As digital technologies and personalized medicine continue to reshape the pharmaceutical landscape, the ethical challenges facing Medical Affairs will become increasingly complex. AI, genetic data, and digital health tools offer tremendous opportunities for improving patient outcomes, but they also require careful ethical oversight. Medical Affairs will remain at the forefront of this evolution, ensuring that patient welfare remains the highest priority, even as the industry adapts to new technologies and rising regulatory scrutiny.

By leading efforts in transparency, ethical decision-making, and compliance monitoring, Medical Affairs can help build a more ethical and patient-centered pharmaceutical industry. The future of healthcare will demand greater digital ethics, more robust patient-centric approaches, and continued commitment to ethical behavior across all company functions. Medical Affairs, as both the ethical and scientific compass, will guide the pharmaceutical industry through these challenges, ensuring that ethics and innovation advance hand in hand.

Case Study: Unethical Practices in the Pharmaceutical Industry—Inspired by the "Pain Hustlers" Movie

In the movie *Pain Hustlers*, the story revolves around unethical practices by a pharmaceutical company that prioritizes profits over patient safety. While fictional, the movie reflects real-world examples of unethical behaviors that have occurred in the industry. One such real-life case closely resembling this narrative involves a pharmaceutical company that engaged in unethical marketing and bribery practices to boost the sales of its opioid product, a fentanyl-based painkiller.

What the Pharma Company Did Wrong

The company engaged in several unethical and noncompliant activities:

- *Bribery of Healthcare Professionals*: The company paid doctors and HCPs to prescribe the medicinal product, often for off-label, non-approved uses. The company held sham speaking events where doctors were paid for promoting the drug, even if the events had no educational purpose.
- *False Claims and Misleading Marketing*: The company misrepresented the safety and efficacy of the medicinal product, minimizing the risks associated with the drug, particularly its highly addictive nature, which contributed to the opioid crisis in the U.S.
- *Targeting Vulnerable Patients*: The company aggressively pushed the medicinal product for patients who didn't necessarily need them, including non-cancer patients, exacerbating the public health crisis around opioid addiction.

Why Did It Happen?

The actions taken by the company were driven by a culture of profit maximization. The company's leadership prioritized financial performance over ethical responsibility and patient welfare. Sales teams were pressured to drive prescriptions at any cost, and incentives were tied to how much medicinal product doctors prescribed, fostering an environment where unethical behavior became the norm.

Several factors contributed to this unethical behavior:

- *Lack of Oversight*: The company lacked strong compliance controls, allowing unethical practices to proliferate without sufficient regulatory checks or internal monitoring.
- *Inadequate Ethics Training*: The company failed to instill an ethical culture within its sales teams, which led to widespread unethical conduct.
- *Pressure for Financial Growth*: The company was under intense pressure to meet sales targets, which led to aggressive and misleading marketing tactics.

How Was It Discovered?
The unethical activities were uncovered through:

- *Whistleblowers*: Several former employees came forward, exposing the company's unethical practices, particularly the fraudulent marketing and bribery schemes.
- *Federal Investigation*: A federal investigation revealed the bribery of doctors and misleading promotional practices. The Department of Justice (DOJ) and other regulatory bodies initiated legal action against the company for its role in contributing to the opioid crisis.

Consequences for the Company
The fallout for the company was significant:

- *Legal Penalties*: The company filed for bankruptcy in 2019 after being ordered to pay $225 million in settlements for its role in the opioid crisis. Several top executives, including the CEO, were convicted of racketeering and fraud.
- *Executive Sentencing*: Several company executives, including founder John Kapoor, were sentenced to prison for their roles in the bribery scheme.
- *Reputational Damage*: The company's unethical practices severely damaged its reputation, leading to its eventual collapse and contributing to broader regulatory scrutiny on pharmaceutical companies, especially concerning opioid products.

Key Takeaways
This case underscores the importance of robust compliance programs, ethical leadership, and a corporate culture that prioritizes patient safety over profits. It also highlights how failure to adhere to ethical and regulatory standards can have catastrophic consequences for a pharmaceutical company, both legally and financially. The Medical Affairs function, with its focus on compliance, ethical marketing, and scientific integrity, is crucial in preventing such unethical practices from occurring.

Chapter 13
The Role of Artificial Intelligence in Healthcare and the Pharmaceutical Industry

Key Highlights

- AI enhances drug discovery, clinical trial optimization, and real-world data analysis.
- Regulatory frameworks (FDA, EMA) adapt to ensure AI safety, efficacy, and transparency.
- Large language models (LLMs) improve diagnostics and decision support but require bias mitigation.
- AI-driven real-world evidence analysis informs regulatory and payer decision-making.
- Ethical and regulatory challenges necessitate continuous monitoring and responsible AI integration.

Artificial intelligence (AI) has emerged as a pivotal technology in transforming healthcare and pharmaceutical industries, offering capabilities that go far beyond traditional digital health tools. The U.S. Food and Drug Administration (FDA), as the primary regulatory body, has engaged deeply with AI's integration in biomedicine, both for its potential to advance health outcomes and the unique challenges it presents. AI's role spans drug discovery, clinical research, and patient care, with applications that significantly influence diagnostics, therapeutic development, and real-time patient monitoring.

AI systems, often machine-learning models, can analyze complex datasets, enabling data-driven decisions that enhance medical product efficacy and safety. The FDA's role in this process is critical; it aims to balance innovative applications of AI with the necessary regulatory rigor to protect patient health and promote public trust. As AI applications expand, especially in healthcare and pharmaceutical research, the FDA's regulatory framework adapts to manage AI's associated risks and opportunities, setting a global precedent for AI governance in health.

A. Krendyukov, *Medical Affairs' Fundamentals: The Next Generation*,
https://doi.org/10.1007/978-3-031-92588-7_13

The FDA's Regulatory Framework for AI in Healthcare

The FDA's journey with AI began in 1995 with its approval of software using neural networks for cervical cancer diagnostics. Although the technology showed promising accuracy, it was ultimately underutilized due to cost-effectiveness challenges, highlighting early issues in implementing AI in healthcare. Since then, the FDA has granted marketing authorizations for nearly 1000 AI-enabled medical devices, with applications predominantly in radiology and cardiology, two fields where AI's diagnostic potential is widely recognized.

The FDA's framework for AI regulation leverages a risk-based approach, ensuring that AI technologies, particularly those with potential patient safety impacts, are reviewed in alignment with their risk profiles. High-risk applications, such as clinical decision support tools used in life-critical diagnoses, receive stringent oversight, whereas lower-risk applications benefit from a more flexible regulatory approach. This methodology ensures the FDA can address the broad and evolving scope of AI without stifling innovation.

Current Challenges in AI Regulation

AI regulation presents unique challenges, especially given the range and diversity of AI technologies, from simple rule-based algorithms to sophisticated machine learning and large language models (LLMs). LLMs, in particular, bring complexities due to their generative capabilities, which can produce unanticipated responses. For instance, an AI system may misinterpret clinical data or generate spurious correlations, risking patient safety. The FDA's response includes developing targeted oversight mechanisms to ensure AI tools meet transparency, safety, and efficacy standards. To support innovation in this area, the FDA initiated a software precertification program, allowing developers to implement regulated updates without requiring additional approvals for each change.

Furthermore, the FDA collaborates internationally to align AI regulations, recognizing that AI-based medical products often serve global markets. Working with bodies like the International Medical Device Regulators Forum (IMDRF), the FDA is contributing to the harmonization of standards for AI, ensuring that international practices support the shared goal of safe and effective AI technologies in healthcare.

AI in Development and Clinical Research

AI-Driven Drug Discovery and Repurposing

In pharmaceutical development, AI has proven invaluable in expediting drug discovery and repurposing. Traditional drug discovery is time-intensive and costly, with a high attrition rate from discovery to market. AI models, trained on multi-omics data (genomics, proteomics, and metabolomics), can identify biological targets, predict molecular interactions, and suggest promising compound candidates. For example, AI algorithms analyze vast datasets to discover drug targets and prioritize lead compounds. This enables drug developers to streamline target selection, increase the success rate of trials, and identify safe therapeutic pathways.

Moreover, AI has facilitated drug repurposing, wherein data from electronic health records, clinical trials, and digital health platforms are analyzed to uncover alternate uses for established drugs. This approach has been pivotal in response to public health emergencies, such as identifying potential COVID-19 treatments through AI-powered analysis of antiviral activity across different datasets.

Enhancing Clinical Trial Design and Patient Safety

AI's influence extends beyond drug discovery, playing a critical role in designing and optimizing clinical trials. Recruitment, adherence, and stratification are three key areas where AI adds value:

- *Participant Recruitment and Representation*: AI can mine large-scale clinical and demographic data to identify suitable trial participants, ensuring a representative sample that reflects the population likely to use the medical product. This is particularly important in addressing gender, race, and ethnic diversity, promoting equitable inclusion.
- *Monitoring Adherence and Retention*: AI can assess adherence via reminders, smart pillboxes, and digital biomarkers, which provide real-time monitoring of patient engagement. Through remote monitoring technologies, trial administrators can track adherence to medication schedules and reduce the risk of attrition, which is critical for the validity of clinical trials.
- *Data Collection and Analysis*: AI enhances clinical trial data management by processing multimodal data, detecting anomalies, and applying imputation tech-

niques to handle missing values. Additionally, AI-driven virtual cohorts simulate clinical records, predicting outcomes and informing clinical trial adjustments that would traditionally rely on extensive real-world patient data.

With AI, adverse event detection becomes faster, allowing earlier intervention to mitigate risks. In this regard, AI can synthesize trial data with post-market surveillance information, thus offering a life-cycle view of drug performance and patient safety.

AI in Clinical Care and Decision Support

The Role of AI in Diagnostics and Patient Monitoring

In clinical care, AI systems primarily aid diagnostics and monitoring. For example, AI has improved diagnostic accuracy in radiology by analyzing imaging data with greater speed and precision than human radiologists, assisting in detecting conditions such as tumors, cardiac abnormalities, and neurodegenerative diseases. Additionally, AI models deployed for remote patient monitoring analyze data from wearable devices, helping clinicians detect deviations in vital signs and providing early warnings of potential complications.

The FDA supports such applications through a lifecycle management approach, requiring that AI tools undergo real-world performance assessments post-deployment. This ensures that AI models continue to perform accurately across varied clinical settings and population subsets. By monitoring AI models in their operational environments, the FDA aims to prevent potential declines in model performance over time, thus safeguarding patient outcomes.

Clinical Decision Support and Large Language Models

Clinical decision support (CDS) tools powered by AI assist clinicians by offering data-driven recommendations. However, while CDS systems can reduce cognitive load and improve clinical decision-making, they introduce risks related to algorithmic bias and over-reliance on AI outputs. LLMs, as a subset of CDS tools, present further complexities. For instance, while LLMs like ChatGPT can interpret unstructured clinical notes or assist in generating clinical summaries, they may produce non-evidence-based recommendations if not carefully calibrated.

The FDA's regulatory approach to CDS tools, particularly for LLMs, emphasizes validation, transparency, and bias mitigation. Rigorous testing and validation in clinical settings are essential to ensure that these tools offer reliable, unbiased support that complements human expertise rather than replacing it.

Challenges and Future Directions

Ethics and Bias in AI Models

The application of AI in healthcare raises ethical considerations, notably around bias and fairness. AI systems trained on historical health data may unintentionally reflect or exacerbate biases present in the original datasets, disproportionately affecting minority groups. The FDA, alongside industry stakeholders, advocates for transparent development practices that prioritize equitable healthcare outcomes. This includes ensuring that AI models are representative of diverse populations and systematically auditing algorithms to identify and correct biases.

Balancing Innovation with Patient Safety

The rapid pace of AI development demands an adaptable regulatory response that can keep up with technological advances without sacrificing patient safety. The FDA's collaboration with other regulatory bodies, such as the European Medicines Agency (EMA) and Health Canada, is pivotal to aligning international practices, ensuring that AI-enabled products are evaluated consistently across global markets. These collaborations allow regulators to share insights, harmonize standards, and develop best practices for AI, thereby fostering an environment where innovation can thrive safely.

The FDA's Total Product Lifecycle (TPLC) approach exemplifies this adaptive strategy, providing a framework to manage AI technologies across their life span. By incorporating pre-market evaluation, post-market monitoring, and recurrent local validation, the TPLC approach aligns AI oversight with the dynamic nature of AI technologies, especially in the context of machine learning models that can evolve post-deployment.

Preparing for the Future of Large Language Models and Generative AI

Generative AI, including LLMs, has garnered significant attention for its potential to revolutionize healthcare by automating administrative tasks, generating clinical documentation, and providing interactive patient support. However, these models also carry risks, particularly concerning unverified outputs and potential "hallucinations." To address these concerns, the FDA emphasizes rigorous regulatory oversight, focusing on clinical validation and risk assessment. Future frameworks may involve specialized evaluation tools for generative AI, ensuring these models perform reliably within clinical settings.

Forward-Looking Perspectives on Artificial Intelligence in Medicine

Predictive Algorithms for Rare Diseases and Personalized Medicine

AI has shown transformative potential in improving patient outcomes, yet several research areas remain underexplored, particularly predictive algorithms for rare diseases and personalized medicine applications. *Rare diseases*—which affect fewer than 200,000 individuals in the United States—pose unique diagnostic and therapeutic challenges due to limited data, diverse presentations, and the small patient populations available for clinical trials. AI models, especially machine learning algorithms, could offer solutions by detecting patterns within limited datasets, facilitating early diagnosis, and identifying novel biomarkers that inform personalized treatment approaches. However, the development of these algorithms requires dedicated datasets and validation protocols that can adapt to the variability inherent in rare diseases.

The complexity of *personalized medicine* also presents opportunities for AI advancements. AI models that predict therapeutic responses and tailor interventions to the individual are critical in moving toward patient-specific care. For example, oncology is at the forefront of AI-driven personalized medicine, where models predict tumor behavior based on genetic profiles, potentially improving treatment efficacy and patient survival. Still, significant gaps remain, particularly in validating these models across diverse demographic groups and specific disease states.

To promote innovation in these areas, the FDA and industry stakeholders could consider:

- *Enhanced Data Sharing Mechanisms*: Rare disease research suffers from fragmented data sources. Establishing secure data-sharing partnerships among academic, regulatory, and private-sector stakeholders can aggregate rare disease data and optimize AI model training. Privacy-preserving AI techniques, like federated learning, could allow institutions to contribute data without compromising patient confidentiality.
- *Development of Specialized Validation Protocols*: Validating AI algorithms for rare diseases and personalized medicine necessitates flexible regulatory frameworks that accommodate small sample sizes and ensure clinical relevance. The FDA may collaborate with research bodies to develop these protocols, fostering a regulatory environment that supports adaptive learning and model refinement.
- *Funding Targeted AI Initiatives*: Additional funding, potentially through public-private partnerships, could support dedicated research initiatives to develop AI tools for rare disease and personalized medicine applications. The FDA could offer grants or incentive programs to encourage academic institutions and start-ups to engage in this complex but high-impact area.

Closing Gaps in AI-Driven Drug Discovery

AI has already impacted drug discovery by identifying potential drug targets, screening compounds, and predicting therapeutic outcomes. However, to fully realize AI's potential, future research must address specific challenges in drug toxicity prediction, pharmacogenomics, and multi-omics integration. Predictive models that can accurately forecast drug toxicity profiles will play a crucial role in preventing adverse events, thus enhancing patient safety and reducing costs associated with late-stage drug development failures.

Pharmacogenomics, or the study of how genes affect a person's response to drugs, holds promise for personalized therapies, particularly in oncology and mental health. AI-driven pharmacogenomic models could optimize dosing regimens and reduce adverse effects; however, current models often require larger, more representative datasets to ensure generalizability across populations.

Additionally, combining multi-omics data sources, such as genomics, transcriptomics, and metabolomics, could offer a comprehensive view of disease mechanisms, but this integration remains technically challenging due to the variability and volume of data.

By supporting future research in these areas, the FDA and other stakeholders can contribute to AI models that enhance therapeutic efficacy, patient safety, and the development of personalized treatment pathways.

Anticipating Future Regulatory Frameworks for Advanced AI Models

Hybrid AI-Human Regulatory Models

Given the increasing sophistication of AI in healthcare, the FDA might consider hybrid AI-human regulatory models to manage advanced AI applications. A hybrid model combines algorithmic evaluations with human oversight, allowing regulators to handle the complex and dynamic nature of AI technologies without compromising safety and efficacy standards.

- *Dynamic Regulatory Approaches for Adaptive Models*: Advanced AI models, especially machine learning systems, can adapt and improve with new data over time. A hybrid AI-human model would involve human regulators working alongside AI algorithms to continually evaluate model updates and adjust oversight requirements based on real-world performance metrics.
- *Risk-Based Assessment Frameworks*: Developing risk-based frameworks tailored to different AI applications could streamline the regulatory process, focusing resources on high-impact, high-risk AI models, such as those used in diagnostics or treatment recommendations. Low-risk applications might undergo

lighter oversight, preserving resources for models with more significant patient safety implications.
- *Localized Validation for Community-Specific Adaptations*: For AI models intended to function in diverse populations, the FDA may explore localized validation schemes. These schemes involve context-specific evaluations of AI models, especially in underserved or diverse communities where demographic variability may affect model performance. By validating AI applications in these localized contexts, the FDA can ensure that models are both clinically relevant and culturally sensitive.

Expanding on Large Language Models and Generative AI Risks

As generative AI models, particularly LLMs, become increasingly prominent in healthcare, the FDA faces the task of evaluating their unique risks and ensuring reliable performance. Unlike traditional models, LLMs generate free-form text based on input prompts, creating the potential for "hallucinations" or unverified outputs that may mislead clinical decision-making.

FDA's Evaluation Strategies for Generative AI Models could include:

- *Controlled Environment Testing*: LLMs could undergo extensive testing in controlled environments before they are integrated into clinical workflows. This would involve analyzing model outputs in hypothetical clinical scenarios to assess accuracy, relevance, and consistency. Controlled testing allows the FDA to identify potential risks, such as biased outputs or inaccuracies, prior to broader deployment.
- *Post-Market Surveillance and Continuous Monitoring*: Since generative AI models may learn and adapt over time, continuous monitoring is essential to detect changes in performance that could impact patient safety. The FDA might implement recurrent post-market surveillance requirements, using automated monitoring tools to track performance changes in real time.
- *Frameworks for Specific Applications*: Given the varied use cases for LLMs, the FDA could develop tailored evaluation frameworks based on application types. For instance, LLMs designed as patient support tools might prioritize patient comprehension and accessibility, while models in clinical decision support must adhere to accuracy and consistency standards.

Exploring AI Scribes and Decision Support in Clinical Workflows

LLMs and other generative AI models show promise as AI scribes, automating documentation tasks in clinical settings. However, the integration of AI scribes requires careful oversight to ensure that these tools enhance clinician efficiency without compromising clinical accuracy or introducing biases.

FDA Guidance on AI Scribes could focus on:

- *Ensuring Complementarity with Human Decision-Making*: AI scribes should serve as supportive tools, not replacements, for clinicians. FDA guidance could mandate transparency mechanisms, where AI-generated outputs are clearly flagged, allowing clinicians to verify accuracy and reduce over-reliance on AI-driven recommendations.
- *Mitigating Cognitive Load and Enhancing Data Accuracy*: By automating administrative tasks, AI scribes can reduce cognitive load for clinicians, allowing them to focus on patient care. However, the FDA's guidance must ensure that AI scribes do not inadvertently omit critical patient details or generate incorrect information. Regular assessments, including clinician feedback, could be instrumental in validating these tools.
- *Establishing Usage Boundaries*: The FDA might define usage boundaries, specifying where AI scribes should be limited to supportive tasks. For instance, while AI scribes can generate summaries and transcriptions, direct clinical decision-making should remain the clinician's responsibility.

Incorporating Visuals to Support Content

To facilitate understanding of complex regulatory structures and AI applications, integrating visuals could enhance reader engagement and comprehension.

Illustrative Diagrams and Flowcharts

- *AI Regulatory Lifecycle Flowchart*: A flowchart detailing the FDA's Total Product Lifecycle (TPLC) approach, including stages such as pre-market review, deployment, and post-market monitoring, would clarify the regulatory journey AI models undergo before reaching clinical use.
- *Comparison Table of Regulatory Approaches*: A comparison table displaying regulatory frameworks from global agencies (e.g., FDA, EMA, Health Canada) could provide context to the reader, showcasing how different jurisdictions handle AI in healthcare.
- *Simplified Model of Adaptive Risk-Based Oversight*: A visual model explaining adaptive oversight based on risk levels can illustrate how the FDA allocates resources to high-impact applications, providing clarity on the benefits of a flexible regulatory framework.

Showcasing AI Applications with Illustrative Examples

Graphs and charts can highlight the adoption trends and impacts of AI in various medical fields:

- *AI in Therapeutic Areas Graph*: A graph displaying the use of AI in different therapeutic areas, such as oncology, cardiology, and mental health, can underscore AI's versatile role across healthcare disciplines.
- *Patient Outcomes over Time*: Data visualizations on patient outcomes achieved with AI, such as diagnostic accuracy improvements or reduced hospital readmission rates, can underscore AI's impact on healthcare quality.

Encouraging Stakeholder Collaboration

The forward trajectory of AI in healthcare depends on collaboration across industry, regulators, and healthcare providers. A united effort will drive the ethical and safe deployment of AI tools while ensuring patient-centered innovation. To achieve this, stakeholders could consider forming consortiums that share best practices, support research, and foster responsible AI development.

Highlighting AI's Potential in Addressing Health Inequities

AI holds promise in closing healthcare gaps, particularly in underserved communities where limited resources have led to disparities in health outcomes. The FDA's leadership in fostering responsible AI development can promote equitable healthcare, ensuring that AI models are accessible, validated, and beneficial across demographic lines. Emphasizing equity as a foundational principle for future AI development will ensure that healthcare innovation serves all patient populations, regardless of geographic, economic, or social status.

AI is poised to transform healthcare and pharmaceuticals, offering advancements across diagnostics, drug development, and patient care. However, this transformation must be supported by stringent, adaptive regulatory frameworks to ensure AI's safe and effective use. The FDA's ongoing efforts, grounded in a lifecycle management approach, exemplify a robust regulatory strategy aimed at balancing the promise of AI with the imperative of patient safety.

The future of AI in healthcare will require concerted collaboration among regulators, healthcare providers, and industry stakeholders. As AI continues to develop, a commitment to transparency, accountability, and continuous improvement will be essential to maximize AI's benefits while mitigating risks. Through vigilant oversight and a focus on patient health outcomes, the FDA and the global regulatory community are working to ensure that AI's integration in healthcare enhances, rather than detracts from, the quality and equity of patient care. At the same time, it offers transformative potential, contingent upon the continued commitment to rigorous oversight, ethical development, and equitable deployment. By embracing a proactive regulatory approach and encouraging multi-stakeholder collaboration, the FDA and global regulators can ensure AI's integration supports safe, effective, and inclusive healthcare advancements on a global scale.

Case Study (Based on Lancet Paper "Randomised Controlled Trials Evaluating Artificial Intelligence in Clinical Practice: A Scoping Review,")

The Transformative Role of AI in Clinical Research

The integration of AI into clinical research is a transformative advancement in modern medicine, offering unprecedented potential to enhance diagnostic precision, optimize patient management, and streamline clinical decision-making. Over the past 5 years, AI applications have rapidly progressed from conceptual frameworks to tools directly influencing clinical outcomes. However, this accelerated growth also demands rigorous validation to ensure these technologies are not only effective but also equitable and safe for diverse patient populations.

Randomized controlled trials (RCTs) remain the gold standard for evaluating the efficacy of medical interventions, including AI systems. A recent scoping review published in *The Lancet Digital Health* comprehensively examined 86 RCTs focused on AI applications in clinical practice. The findings highlight the substantial progress AI has made across specialties while also emphasizing critical gaps and challenges.

The Current Landscape of AI in Clinical Practice

The reviewed RCTs reveal an expanding interest in AI applications across a diverse range of medical specialties and geographic locations. Gastroenterology and radiology emerged as the dominant fields, collectively accounting for 56% of the trials. Deep learning algorithms, particularly those used in video-based medical imaging for endoscopic procedures, demonstrated promising diagnostic capabilities. For instance, AI systems significantly improved adenoma detection rates during colonoscopies, addressing critical gaps in cancer screening.

Geographically, the United States and China were leaders in conducting AI-focused trials, collectively contributing to 61% of the reviewed studies. However, these trials were predominantly single-center studies, raising concerns about their generalizability. Furthermore, the concentration of trials within specific specialties and regions underscores the need for a broader, more inclusive research approach that encompasses underrepresented areas such as primary care and low-resource settings.

Efficacy and Performance: What the Evidence Shows

Approximately 81% of the RCTs reported favorable outcomes for their primary endpoints, with diagnostic yield and performance being the most commonly assessed measures. AI-assisted clinicians demonstrated statistically significant improvements compared to unassisted clinicians in 71% of trials, particularly in specialties such as gastroenterology and radiology.

Beyond diagnostic accuracy, AI systems showed potential in enhancing care management and patient behavior. For example, AI-driven tools for insulin dosing improved time-in-range glucose levels in diabetic patients, while prediction systems for diabetic retinopathy increased referral adherence. These findings underscore AI's capacity to augment traditional clinical workflows and empower patient-centered care.

Despite these successes, it is important to temper enthusiasm with critical appraisal. The predominance of narrowly defined endpoints, such as diagnostic accuracy, limits the assessment of AI's true clinical impact. For instance, while AI-assisted colonoscopy improved polyp detection rates, its effect on detecting clinically significant advanced adenomas was less clear. Future trials must prioritize outcomes that directly correlate with patient health and well-being.

Challenges and Limitations

While the scoping review highlights the potential of AI, it also underscores significant challenges that must be addressed. The heavy reliance on single-center trials, representing 63% of studies, limits the generalizability of findings to broader patient populations and healthcare settings. Furthermore, demographic reporting remains insufficient, with race or ethnicity data provided in only 22 trials, predominantly those conducted in the United States.

The lack of adherence to standardized reporting guidelines, such as the CONSORT-AI extension, further hampers transparency and reproducibility. Only 19% of trials published after 2021 referenced these guidelines, despite their critical role in ensuring methodological rigor. This gap highlights the need for greater awareness and implementation of reporting standards to improve the quality of AI research.

Another pressing issue is the operational inefficiency associated with some AI systems. While 35% of trials reported a significant reduction in operational time, 25% noted increased time requirements, particularly in gastroenterology applications. These mixed results emphasize the importance of evaluating not only AI's clinical efficacy but also its impact on workflow and resource allocation.

Operational Efficiency and Real-World Integration

Operational efficiency is a key determinant of AI's viability in clinical practice. For instance, AI systems evaluated in radiology and ophthalmology consistently reduced operational time, highlighting their potential to streamline workflows in high-demand specialties. Conversely, increased time requirements in certain gastroenterology trials may reflect the need for better integration of AI tools into existing clinical processes.

Successful adoption of AI depends on its ability to complement, rather than complicate, clinical workflows. This requires robust training programs for clinicians, intuitive user interfaces, and systems capable of delivering actionable insights in real time. Furthermore, cost-effectiveness analyses should be integral to future RCTs, ensuring that AI solutions provide tangible value to healthcare systems.

Future Directions and Recommendations

To realize AI's full potential in clinical research, a multifaceted approach is required. First, future RCTs should prioritize clinically meaningful endpoints that directly impact patient outcomes, such as symptom relief, quality of life, and long-term survival. Emphasizing patient-centered metrics will provide a more comprehensive understanding of AI's benefits and limitations.

Second, greater investment in multicenter trials is essential to enhance the generalizability of findings. Collaboration across institutions and countries will help ensure that AI systems are validated in diverse populations and healthcare environments. This approach is particularly important for addressing health disparities and ensuring equitable access to AI-driven innovations.

Third, adherence to reporting standards such as CONSORT-AI should be mandatory for all AI-focused RCTs. Transparent reporting will facilitate critical evaluation and replication, fostering trust in AI research. Additionally, regulatory harmonization across jurisdictions will streamline the approval process for AI-based interventions, reducing barriers to implementation.

Fourth, federated learning presents an innovative approach to overcome data-sharing barriers in clinical research. By enabling AI models to learn across institutions without compromising patient privacy, federated learning fosters collaboration and ensures diverse representation in training datasets. This approach has been successfully demonstrated in developing models for rare disease diagnosis and imaging-based tools for cancer detection. As privacy regulations such as GDPR tighten, federated learning will play a crucial role in the sustainable advancement of AI systems.

Fifth, real-world evidence (RWE) is increasingly recognized as a complement to RCTs in assessing treatment effectiveness and safety. AI systems can analyze vast amounts of real-world data from electronic health records, registries, and wearable devices to identify adverse events and long-term outcomes. For instance, AI algorithms have been used to detect rare but critical side effects of newly approved drugs faster than traditional pharmacovigilance methods. By integrating RWE with trial data, AI offers a holistic view of therapeutic impact, informing both clinical practice and regulatory decisions.

Finally, future research should explore the integration of multimodal AI systems that incorporate diverse data sources, such as electronic health records, imaging, and genomic data. These systems hold the promise of delivering more personalized and context-aware clinical recommendations, further advancing the precision medicine paradigm.

AI has the potential to revolutionize clinical research and practice, as evidenced by the promising results of recent RCTs. However, this early success must be tempered by a recognition of the field's challenges, including issues of generalizability, transparency, and operational efficiency. By addressing these gaps and prioritizing patient-centered outcomes, the medical community can harness AI's transformative power to improve healthcare delivery and outcomes.

As AI continues to evolve, ongoing collaboration between clinicians, researchers, and policymakers will be essential to ensure its responsible and effective integration into clinical practice. The future of AI in medicine is undeniably bright, but it requires a steadfast commitment to evidence-based innovation and ethical stewardship.

Case Study: Pioneering AI in the Medicinal Product Lifecycle—Insights from EMA's Reflection Paper

The EMA has recently published a reflection paper on the use of AI in the medicinal product lifecycle. This comprehensive document serves as a critical milestone in integrating AI within pharmaceutical development, approval, and monitoring processes. In this section, we delve into the significance of the EMA's guidance, its implications for the pharmaceutical industry, and the transformative potential it holds for the future of healthcare.

The Digital Transformation of Pharma

AI has emerged as a cornerstone of innovation across industries, and healthcare is no exception. With applications ranging from drug discovery to post-marketing surveillance, AI's potential to revolutionize the medicinal product lifecycle is unprecedented. Recognizing this potential, the EMA's reflection paper addresses how AI can be harnessed effectively and responsibly while ensuring patient safety, regulatory compliance, and ethical integrity. This forward-looking framework underscores the importance of regulatory clarity in fostering innovation while safeguarding public health.

Key Highlights of the EMA's Reflection Paper

The reflection paper provides detailed insights into AI's applications, challenges, and governance within the medicinal product lifecycle. It covers critical areas such as risk management, regulatory interactions, technical considerations, and ethical implications.

Definitions and Scope

The EMA defines AI broadly, encompassing machine learning (ML) and other adaptive technologies that exhibit varying levels of autonomy. The scope of the guidance includes all stages of the medicinal product lifecycle, from drug discovery and clinical trials to post-authorization activities.

General Considerations

The EMA emphasizes a risk-based approach to developing and deploying AI systems. High-risk applications, such as those affecting patient safety or regulatory decision-making, require rigorous validation and oversight. Developers must ensure that AI systems are fit for purpose, transparent, and adhere to ethical standards.

AI Applications Across the Medicinal Product Lifecycle

The reflection paper identifies several key areas where AI can enhance efficiency and accuracy:

- *Drug Discovery*: AI enables rapid identification of potential drug candidates by analyzing large datasets. This accelerates early-stage research and reduces costs.
- *Non-Clinical Development*: AI can refine preclinical studies, improving human translatability and reducing reliance on animal testing.
- *Clinical Trials*: From trial design to data analysis, AI enhances efficiency, reduces bias, and supports decentralized trials.
- *Precision Medicine*: AI facilitates personalized treatment approaches by analyzing patient-specific data, biomarkers, and genetic profiles.

- *Manufacturing*: AI optimizes production processes, ensuring consistency and quality in drug manufacturing.
- *Post-Authorization Activities*: AI supports pharmacovigilance, signal detection, and post-marketing surveillance by identifying adverse events and monitoring product efficacy.

Technical and Ethical Considerations

The EMA's paper provides specific guidance on technical aspects such as data management, model validation, and performance monitoring. It highlights the importance of:

- *Data Integrity*: Ensuring high-quality, representative datasets to mitigate bias.
- *Model Transparency*: Promoting explainability and accountability, particularly for "black-box" models.
- *Continuous Monitoring*: Implementing mechanisms to detect model drift and maintain performance over time.

Ethical principles underpin the EMA's guidance, emphasizing human oversight, non-discrimination, and patient-centricity. Developers are encouraged to adopt the European Commission's guidelines for trustworthy AI, which prioritize transparency, fairness, and societal well-being.

Implications for the Pharmaceutical Industry

The EMA's reflection paper has far-reaching implications for pharmaceutical companies, regulatory bodies, and other stakeholders. Key considerations include:

- *Accelerating Innovation*: By providing clear regulatory pathways, the guidance fosters innovation and encourages the adoption of AI technologies.
- *Enhancing Efficiency*: AI streamlines complex processes, reduces development timelines, and optimizes resource allocation.
- *Building Trust*: The emphasis on transparency and ethical governance strength ens stakeholder confidence in AI-driven solutions.
- *Global Harmonization*: Aligning with international standards facilitates cross-border collaboration and market access.

Challenges and Opportunities

While the potential of AI is immense, the EMA's guidance acknowledges several challenges:

- *Data Bias*: Ensuring datasets are diverse and representative remains a significant challenge.
- *Regulatory Compliance*: Navigating complex regulatory requirements demands robust documentation and validation processes.
- *Transparency*: Communicating complex AI mechanisms to non-technical stakeholders requires effective tools and frameworks.

Opportunities

- *Advancing Data Science*: The need for high-quality datasets drives innovation in data collection, curation, and augmentation.

- *Collaborative Development*: Multidisciplinary collaboration between data scientists, clinicians, and regulatory experts fosters innovation.
- *Improving Patient Outcomes*: AI-driven insights enable more effective treatments, ultimately benefiting patients.

A Collaborative Path Forward

The EMA's reflection paper marks a pivotal moment in the integration of AI within the pharmaceutical industry. By addressing technical, ethical, and regulatory considerations, it provides a robust framework for leveraging AI's transformative potential. For the pharmaceutical sector, this represents an opportunity to innovate responsibly, enhance patient care, and navigate the complexities of an evolving regulatory landscape.

As AI technologies continue to advance, ongoing collaboration among regulators, industry stakeholders, and academia will be essential. The EMA's proactive approach sets a precedent for fostering innovation while safeguarding public health, paving the way for a future where AI is seamlessly integrated into the fabric of healthcare.

Case Study: Overarching Perspectives on FDA and EMA Approaches to AI in Medicinal Product Lifecycle (EMA Reflection Paper 2024 and FDA New Guidelines 2025), Table 13.1

The integration of AI in healthcare has ushered in transformative changes across the pharmaceutical and medical device industries. Both the FDA and EMA have issued comprehensive guidelines—the FDA's *Artificial Intelligence-Enabled Device Software Functions: Lifecycle Management and Marketing Submission Recommendations* and the EMA's *Reflection Paper on the Use of Artificial Intelligence in the Medicinal Product Lifecycle*. These documents represent pivotal milestones in defining regulatory expectations for AI technologies, fostering innovation, and ensuring patient safety.

Shared Regulatory Principles

Both the FDA and EMA recognize the transformative potential of AI and its role in advancing healthcare. Their guidelines reflect a commitment to fostering innovation while maintaining stringent safety and efficacy standards. Several overarching principles are shared between the two agencies:

Lifecycle Approach to AI Regulation

Both agencies emphasize a Total Product Lifecycle (TPLC) approach to managing AI systems. This framework ensures that AI technologies are assessed, monitored, and refined throughout their lifecycle, from development and validation to deployment and post-market monitoring. By addressing risks proactively, this approach fosters trust and reliability.

Risk-Based Assessment

A risk-based framework underpins the guidance from both agencies. They classify AI applications based on their potential impact on patient safety and regulatory decision-making. High-risk systems, such as those used for clinical trial decision-making or patient-specific dosing recommendations, are subject to more rigorous validation and oversight.

Table 13.1 FDA and EMA AI regulatory comparison

Aspect	*FDA*	*EMA*
Lifecycle approach to AI regulation	Both emphasize total product lifecycle management.	
Risk-based assessment	Risk frameworks focus on patient safety and decision-making impact.	
Data integrity and bias mitigation	Both require representative datasets to reduce bias.	
Transparency and Explainability	Transparency and explainability are prioritized in high-risk applications.	
Continuous monitoring and performance validation	Performance monitoring and drift detection are critical for sustained safety	
Scope of guidance	Focuses on AI-Device Software Functions and detailed marketing submission requirements.	Broader perspective covering the entire medicinal product lifecycle.
Regulatory interactions	Emphasizes premarket submissions and documentation of AI systems.	Encourages early interaction with the innovation task force and seeking advice.
Ethical and societal considerations	Primarily focuses on technical aspects with limited ethical emphasis.	Strong emphasis on ethics, societal well-being, and human-centric design.
Transparency expectations	Transparency is tied to documentation and traceability for regulatory reviews.	Includes explainability for patients and healthcare providers.
Post-market monitoring	Requires structured mechanisms for post-market drift detection.	Integrates AI into pharmacovigilance frameworks, leveraging real-world evidence.

Data Integrity and Bias Mitigation

Data integrity and diversity are central to both the FDA and EMA's recommendations. They stress the importance of representative datasets to prevent bias and ensure equitable outcomes. Strategies such as over-sampling rare populations and leveraging diverse datasets are encouraged to enhance model generalizability.

Transparency and Explainability

Transparency in AI models is a shared priority. Both agencies advocate for explainability, particularly in high-risk applications, to enable users and regulators to understand the rationale behind AI-driven decisions. The use of explainability metrics, such as SHAP and LIME, is recommended to elucidate model behavior.

Continuous Monitoring and Performance Validation

The FDA and EMA emphasize the importance of ongoing performance monitoring to detect model drift and ensure sustained effectiveness. This includes post-market surveillance and mechanisms to refine AI models as new data becomes available.

Applications Across the Medicinal Product Lifecycle

Both agencies outline similar applications of AI across the medicinal product lifecycle, underscoring its transformative potential for drug discovery and preclinical development, clinical trials, precision medicine, manufacturing, post-marketing surveillance.

Divergent Approaches: FDA vs. EMA

While the FDA and EMA share many principles, their approaches diverge in notable ways, reflecting differences in regulatory philosophies, priorities, and regional contexts (Table 13.1).

Scope of Guidance

- *FDA*: The FDA's guidance is focused on AI-enabled device software functions (AI-DSFs), reflecting its broader mandate to oversee medical devices. The guidance is highly detailed, providing specific recommendations for marketing submissions and lifecycle management.
- *EMA*: The EMA's reflection paper adopts a broader perspective, addressing AI applications across the entire medicinal product lifecycle, from drug discovery to post-authorization activities. This holistic approach reflects the EMA's focus on medicines regulation.

Regulatory Interactions

- *FDA*: The FDA emphasizes premarket submissions, requiring detailed documentation of AI systems' design, validation, and performance metrics. It also outlines a process for modifications to AI systems post-approval.
- *EMA*: The EMA highlights early regulatory interactions, encouraging developers to engage with its Innovation Task Force and seek scientific advice to align AI applications with regulatory expectations. This approach fosters collaboration and anticipates potential challenges.

Ethical and Societal Considerations

- *FDA*: The FDA's guidance primarily focuses on technical and operational aspects of AI, with limited emphasis on broader ethical issues.
- *EMA*: The EMA's paper places a stronger emphasis on ethical principles, including non-discrimination, societal well-being, and the human-centric design of AI systems. It aligns with the European Union's broader regulatory framework, such as the GDPR and AI Act.

Transparency Expectations

- *FDA*: The FDA's focus on transparency is pragmatic, prioritizing documentation and traceability to support regulatory review and user trust.
- *EMA*: The EMA's guidance extends transparency requirements to include explainability for patients and healthcare providers, reflecting its commitment to equitable access and understanding.

Post-Market Monitoring

- *FDA*: The FDA emphasizes structured post-market monitoring, requiring manufacturers to establish mechanisms for detecting and mitigating model drift.

- *EMA*: While the EMA also underscores the importance of post-market monitoring, its guidance highlights the integration of AI systems within pharmacovigilance frameworks, leveraging real-world evidence to refine regulatory decisions.

Toward Harmonized AI Regulation

The FDA and EMA's guidance on AI represents a significant step toward integrating cutting-edge technologies into the pharmaceutical and medical device industries. While their approaches differ in scope and emphasis, both agencies share a commitment to fostering innovation, ensuring patient safety, and maintaining ethical integrity.

For stakeholders, understanding these guidelines is crucial to navigating the evolving regulatory landscape. By embracing the shared principles and addressing regional nuances, the industry can harness AI's transformative potential to improve patient outcomes and advance global healthcare.

The path forward lies in continued collaboration between regulators, industry, and academia. Harmonizing AI regulations across regions will be instrumental in unlocking its full potential, paving the way for a future where AI is seamlessly integrated into every facet of healthcare.

Case Study: Advancing Hospital Performance Through Strategic AI Adoption

The intersection of AI and healthcare is a burgeoning field that offers transformative potential for hospital operations and patient care. The recently published study by Phuoc Pham et al., titled "Determinants and Performance Outcomes of AI Adoption: Evidence from U.S. Hospitals," provides a robust empirical analysis of this potential. This case study aims to contextualize and expand on the findings, highlighting the critical determinants and performance implications of AI adoption in hospitals.

AI in Healthcare: Opportunities and Challenges

AI technologies, encompassing machine learning, natural language processing, and deep learning, have emerged as vital tools for healthcare organizations. These technologies enable the processing and analysis of large datasets to enhance diagnostics, optimize administrative workflows, and improve patient outcomes. According to Accenture's analysis, clinical AI applications could generate $150 billion in annual savings for the U.S. healthcare system by 2026. Despite such promise, the adoption of AI in healthcare remains uneven, with numerous organizational and systemic barriers slowing its integration.

A recently published example in the US hospital provides a nuanced exploration of the determinants influencing AI adoption, emphasizing the importance of organizational factors such as market share and average length of stay. Furthermore, the research sheds light on the long-term benefits of AI for hospital financial and operational performance, offering actionable insights for healthcare leaders.

Key Determinants of AI Adoption

The study identifies two primary determinants of AI adoption: market share and the average length of stay (ALOS). These factors underline the role of organizational capacity and operational complexity in influencing a hospital's readiness to implement AI technologies.

Market Share

Hospitals with a larger market share are better positioned to adopt AI technologies. This finding aligns with the Organizational Information Processing Theory (OIPT), which posits that organizations with greater resources and complexity require advanced information-processing capabilities. Larger hospitals benefit from economies of scale, broader physician networks, and higher patient volumes, enabling them to allocate significant financial and human resources to AI implementation. These institutions also have better access to capital and more extensive reimbursement mechanisms, further facilitating the integration of advanced technologies.

Average Length of Stay

In contrast to market share, the study finds a negative relationship between ALOS and AI adoption. While longer patient stays introduce operational complexities that could benefit from AI-driven efficiencies, hospitals with extended ALOS may face heightened uncertainties that complicate the adoption process. This finding highlights the need for targeted strategies to address specific operational challenges associated with ALOS, such as predictive analytics for patient discharge planning and resource allocation.

Performance Outcomes of AI Adoption

The study provides compelling evidence of the positive impact of AI adoption on hospital performance. The study's rigorous analysis demonstrates improvements across key financial and operational metrics, underscoring the transformative potential of AI in healthcare.

Financial Performance

AI adoption is significantly associated with increased outpatient and inpatient revenues. Hospitals leveraging AI can optimize resource utilization, enhance service delivery, and improve billing accuracy. For example, AI-driven systems like Computer-Assisted Physician Documentation (CAPD) have demonstrated substantial cost savings by reducing transcription expenses and improving the documentation of patient conditions.

Although the financial benefits of AI may not be immediately apparent due to initial investment costs and the technology's learning curve, the long-term gains are evident. Hospitals investing in AI can expect sustained revenue growth as they refine their use of these technologies.

Operational Performance

Operational improvements associated with AI adoption include enhanced productivity and occupancy rates. By automating routine tasks and streamlining workflows, AI reduces administrative burdens on healthcare staff, allowing them to focus on patient care. The study highlights how AI enables better demand forecasting, resource allocation, and interdepartmental coordination, ultimately improving efficiency.

These operational gains are particularly crucial in the current healthcare landscape, where hospitals face increasing pressure to deliver high-quality care with limited resources. AI technologies provide a scalable solution to these challenges, enabling hospitals to achieve better outcomes while optimizing costs.

Implications for Healthcare Leaders

These findings offer several practical implications for hospital administrators, policymakers, and other stakeholders in the healthcare ecosystem.

Strategic Investments in AI

Hospitals with substantial market share should leverage their resources to invest in AI technologies. These investments should focus on areas where AI can deliver the greatest impact, such as patient flow management, diagnostic accuracy, and revenue cycle optimization. Smaller hospitals, while facing resource constraints, can explore partnerships and collaborative models to access AI capabilities.

Addressing Barriers to Adoption

The study underscores the importance of overcoming barriers to AI adoption, including high implementation costs, limited in-house expertise, and data challenges. Hospitals should prioritize building internal capabilities through training programs and partnerships with technology providers. Policymakers can also play a role by offering financial incentives and regulatory frameworks that support AI adoption.

Fostering a Data-Driven Culture

Successful AI integration requires a cultural shift within healthcare organizations. Hospitals must embrace a data-driven approach to decision-making, fostering collaboration across departments and encouraging the adoption of new technologies. Leadership commitment is essential in driving this cultural transformation, ensuring that AI initiatives align with broader organizational goals.

Leveraging AI for Resilience

The COVID-19 pandemic has highlighted the importance of resilience in healthcare systems. AI technologies can enhance hospital resilience by enabling real-time monitoring, predictive analytics, and adaptive decision-making. These capabilities are critical for managing disruptions and ensuring continuity of care during crises.

Future Directions and Research Opportunities

While this example provides valuable insights, several areas warrant further exploration:

- *Longitudinal Analysis*: Future research should examine the long-term impacts of AI adoption on hospital performance, considering factors such as technology evolution and changing healthcare dynamics.
- *Comparative Studies*: Comparative analyses across different healthcare systems and geographic regions can provide a broader understanding of the factors influencing AI adoption.
- *Patient-Centered Outcomes*: While the study focuses on financial and operational metrics, additional research is needed to evaluate the impact of AI on patient outcomes, such as satisfaction, safety, and quality of care.
- *Integration with Other Technologies*: Exploring the synergies between AI and other healthcare technologies, such as telemedicine and IoT, can uncover new opportunities for innovation.
- *Ethical and Regulatory Considerations*: As AI adoption accelerates, addressing ethical concerns and ensuring compliance with regulatory standards will be critical. Research should explore frameworks for responsible AI implementation in healthcare.

Summary: The adoption of AI technologies in hospitals represents a paradigm shift in healthcare delivery. As Pham et al. demonstrate, AI has the potential to significantly enhance hospital performance, driving improvements in financial and operational outcomes. By strategically addressing the determinants of AI adoption and leveraging its transformative capabilities, healthcare leaders can unlock new opportunities for innovation and excellence.

This case study underscores the importance of evidence-based decision-making in guiding AI adoption strategies. As the healthcare landscape continues to evolve, the insights provided by this study offer a roadmap for harnessing the power of AI to advance hospital performance and improve patient care (some of the terms are listed in Table 13.2). The journey toward widespread AI adoption is complex but ultimately rewarding, promising a future where technology and human expertise converge to deliver unparalleled healthcare outcomes.

Table 13.2 Some practical terms in generative AI

Term	Definition
LLM (large language model)	Advanced AI systems trained on massive text datasets
Transformers	Neural network architecture that revolutionized AI
Prompt engineering	The art of crafting effective instructions for AI models
Fine-tuning	Customizing pre-trained models for specific tasks
Embeddings	Numerical representations of text/images
RAG (retrieval augmented generation)	Combining external knowledge with AI generation
Tokens	Units of text processing in AI models
Hallucination	AI generating false but plausible information
Zero-shot learning	AI performing tasks without specific training
Chain-of-thought	Step-by-step reasoning in AI responses
Context window	Maximum text length AI can process
Temperature	Controls AI output randomness
Generative adversarial networks (GANs)	A class of machine learning frameworks where two neural networks contest each other to generate new, synthetic instances of data
Diffusion models	Probabilistic models that generate data by reversing a diffusion process, often used in image and audio generation
Neural networks	Computing systems inspired by the biological neural networks that constitute animal brains, capable of pattern recognition and data classification
Deep learning	A subset of machine learning involving neural networks with multiple layers, enabling the modeling of complex patterns in large datasets
Natural language processing (NLP)	A field of AI that focuses on the interaction between computers and humans through natural language

(continued)

Table 13.2 (continued)

Term	Definition
Reinforcement learning	An area of machine learning where an agent learns to make decisions by performing actions and receiving feedback from the environment
Autoencoders	A type of artificial neural network used to learn efficient codings of unlabeled data, primarily for dimensionality reduction
Transfer learning	A machine learning technique where a model developed for a particular task is reused as the starting point for a model on a second task
Synthetic data	Data that is artificially generated rather than obtained by direct measurement, used to train AI models when real data is scarce or sensitive
Self-supervised learning	A type of learning where the data provides the supervision, allowing models to learn representations without explicit labels
Attention mechanism	A component of neural networks that enables the model to focus on specific parts of the input sequence, improving performance in tasks like translation

Further Reading

ABPI 2024 Code of Practice

EMA Reflection paper on the use of Artificial Intelligence (AI) in the medicinal product lifecycle, 2024 https://www.ema.europa.eu/system/files/documents/scientific-guideline/reflection-paper-use-artificial-intelligence-ai-medicinalproduct-lifecycle-en.pdf

FDA artificial intelligence-enabled device software functions: lifecycle management and marketing submission recommendations, Draft Guidance for Industry and Food and Drug Administration Staff, Jan 2025 https://www.fda.gov/regulatoryinformation/search-fda-guidance-documents/artificial-intelligence-enabled-device-software-functions-lifecycle-management-andmarketing

Han R et al (2024) Randomised controlled trials evaluating artificial intelligence in clinical practice: a scoping review. Lancet Digit Health 6:e367–e73l. https://doi.org/10.1016/S2589-7500(24)00047-5 www.thelancet.com/digital-health Vol 6 May 2024

Henderson RH, French D, Stewart E et al (2023) Delivering the precision oncology paradigm: reduced R&D costs and greater return on investment through a companion diagnostic informed precision oncology medicines approach. J Pharm Policy and Pract 16:84. https://doi.org/10.1186/s40545-023-00590-9

Iglesias I et al (2024) Understanding the National Healthcare Ecosystem to position medical affairs as a strategic element: lessons Learned from AstraZeneca Spain. Pharm Med 39. https://doi.org/10.1007/s40290-024-00542-x

IQVIA Global Oncology Trends 2024: Outlook to 2028

Krendyukov A (2019) Innovative oncology products: time to revisit the strategy development? ESMO Open 4(5):e000571. https://doi.org/10.1136/esmoopen-2019-000571

Krendyukov A (2020) Early access provision for innovative medicinal products in oncology: challenges and opportunities. Front Oncol 10. https://doi.org/10.3389/fonc.2020.01604

Krendyukov A, Gattu S (2020) Critical factors shaping strategy development of an innovative medicine in oncology. Pharmaceut Med. 34(2):103–112. https://doi.org/10.1007/s40290-020-00328-x

Krendyukov A, Lerchenmueller M. Challenges and opportunities to the implementation of adaptive design in phase III oncology trials: results from a cross-sectional analysis. ESMO 2023 annual congress, October 2023a. https://doi.org/10.1016/j.annonc.2023.09.1895

Krendyukov A, Lerchenmueller M. Adaptive design in clinical research: opportunities and challenges to improve the information's accessibility and transparency. Open Innovation in Science Research Conference, May 2023b

Krendyukov A, Nasy D (2020) Evolving communication with healthcare professionals in the pharmaceutical space: current trends and future perspectives. Pharmaceut Med 34(4):247–256. https://doi.org/10.1007/s40290-020-00341-0

A. Krendyukov, *Medical Affairs' Fundamentals: The Next Generation*, https://doi.org/10.1007/978-3-031-92588-7

Krendyukov A, Nasy D (2022) Medical affairs and innovative medicinal product strategy development. Pharm Med 36. https://doi.org/10.1007/s40290-022-00424-0

Krendyukov A, Singhvi S, Zabransky M (2021) Value of adaptive trials and surrogate endpoints for clinical decision-making in rare cancers. Front Oncol 11. https://doi.org/10.3389/fonc.2021.636561

Krendyukov A, Bakowska M, Schiller D, Singhvi S (2024) Current understanding, knowledge, and perception of biosimilars in a changing landscape of regulatory requirements. Gener Biosimilars Initiative J (GaBI Journal) 13(1):27–33. https://doi.org/10.5639/gabij.2024.1301.005

Pham P et al (2024) Determinants and performance outcomes of artificial intelligence adoption: evidence from U.S. Hospitals Journal of Business Research 172:114402. https://doi.org/10.1016/j.jbusres.2023.114402

Porter ME, Lee TH (2013) The strategy that will fix health care: providers must lead the way in making value the overarching goal. Harvard Bus Rev 91(10):50–67

GPSR Compliance

The European Union's (EU) General Product Safety Regulation (GPSR) is a set of rules that requires consumer products to be safe and our obligations to ensure this.

If you have any concerns about our products, you can contact us on ProductSafety@springernature.com

In case Publisher is established outside the EU, the EU authorized representative is:

Springer Nature Customer Service Center GmbH
Europaplatz 3
69115 Heidelberg, Germany

Batch number: 10370708

Printed by Printforce, the Netherlands